ELEMENTS OF
PHYSICAL CHEMISTRY

ELEMENTS OF PHYSICAL CHEMISTRY

Arun Rastogi

ANMOL PUBLICATIONS PVT. LTD.
NEW DELHI-110 002 (INDIA)

ANMOL PUBLICATIONS PVT. LTD.

Regd. Office: 4360/4, Ansari Road, Daryaganj,
New Delhi-110002 (India)
Tel.: 23278000, 23261597, 23286875, 23255577
Fax: 91-11-23280289
Email: anmolpub@gmail.com
Visit us at: www.anmolpublications.com

Branch Office: No. 1015, Ist Main Road, BSK IIIrd Stage
IIIrd Phase, IIIrd Block, Bangalore-560 085 (India)
Tel.: 080-41723429 • Fax: 080-26723604
Email: anmolpublicationsbangalore@gmail.com

Elements of Physical Chemistry

First Edition, 2010

ISBN 978-81-261-4490-7

PRINTED IN INDIA

Printed at Mehra Offset Press, Delhi.

Contents

Preface

The basis of physical chemistry is the ability to solve numerical problems. It is generally agreed that it is not with inorganic and organic chemistry that first-year and preliminary-year undergraduate students have the greatest difficulty, but instead the numerical problem-solving aspect of physical chemistry. The global trend of below-average marks in physical chemistry in first-year and preliminary- year chemistry papers needs to be addressed. The preparation of this book has been made much easier by the improvements in the technology of book production. This has resulted in the production of much more colourfully and attractive of general and introductory chemistry. This would not be a problem if the basic principles of chemistry were still clearly identifiable. Unfortunately, often this is not the case and the principles, even when well described, are lost beneath a wealth of factually unconnected data, that is unnecessary for the student to learn.

The basis of physical chemistry is the ability to solve numerical problems. It is generally agreed that it is not with inorganic and organic chemistry that first-year and preliminary-year undergraduate students have the greatest difficulty, but instead the numerical problem-solving aspect of physical chemistry. Those students who want to learn physical chemistry are invited to expand their vision by more specialized works. I furthermore hope that it might also be of interest to all of those who have already been touched, directly or indirectly, by this beautiful and highly creative area of modern science and who would like to learn more about its appeal and promise. It's ideal for the one-semester physical chemistry course, providing an introduction to the essentials of the subject without too much math.

Author

Chapter 1

Behaviour of Gases

KINETIC THEORY OF GASES

The phases (physical states) in which a pure substance may exist. For most substances there are three phases — solid (s), liquid (l), and gas (g). The gas phase is the simplest of the three to deal with theoretically, primarily because the molecules behave completely chaotically and therefore give rise to simply-formulated average properties. The solid phase is also relatively easily dealt with, because it is so highly organized.

Gases and their properties play an ubiquitous and critical role in our daily lives. Life is supported by the oxygen of the air, which reacts with glucose in biochemical combustion. Living systems have evolved complex molecular architecture to absorb oxygen from the air; carry it to the cells; deliver it to the cells; transport it within the cell; and supervise its multistep, controlled reaction with glucose.

It is the rapid expansion of gases produced in the explosive combustion of gasoline that performs the work of the internal combustion engine. Expansion and contraction of gases in response to temperature and pressure changes is at the basis of continually changing weather patterns. Clearly an understanding of the properties, behaviour, and chemical reactivity of gases is an important foundation for developing an appreciation of complex biological and technological systems.

Following a discussion of the ideal gas law, a simple mathematical model for gas behaviour, we discuss the Kinetic

Molecular Theory of gases, which establishes a link between molecular motion and temperature.

PROPERTIES OF GASES

All gases have observable properties in common:

- Gases flow readily from one space to another and occupy all available space.
- Gases assume the shapes of their containers.
- Gases are readily compressible. By exerting force, we can cause a gas to occupy a smaller volume.
- Two or more gases form homogeneous mixtures (solutions) in all proportions. An example is air, a mixture of primarily nitrogen and oxygen with small amounts of other gases.
- Gases diffuse rapidly. This means that a gas can move across a large space in a relatively short time. Examples are the odors of perfume and skunk.

The state of a gas is described by the values of 4 macroscopic variables: the volume (V) occupied by the gas; the temperature (T) of the gas; the pressure (P) that the gas exerts on the walls of its container; and the amount of gas in moles (n). The values of these quantities are related by a very simple equation called the ideal gas law. We will discuss this equation after a brief word on pressure.

PRESSURE.

Pressure is defined as force per unit area. We illustrate the concept of pressure with an example.

Example: The bottom of each foot of a 150-pound man has a surface area of about 24 square inches. When the man stands on a bathroom scale, what pressure is exerted by his feet?

Solution: The force exerted by the feet is the man's weight, 150 lb. This is distributed over the area of the bottoms of both feet, or 48 square inches.

The pressure is then,

$$P = F/A = 150\ lb/48\ in^2 = 3.1\ lb/in^2$$

If the man stands on the scale with only one foot, the

force is the same, but the pressure is doubled because the area is cut in half. Gases exert pressure on the walls of their containers. Since a gas uniformly occupies a container, it exerts the same pressure on all the walls. The gases that make up the atmosphere exert pressure on the surface of the earth.

This pressure, called the atmospheric pressure, can be measured using a barometer. The left side of the figure shows a beaker or cup containing mercury. A glass tube of length 1 m and cross-sectional area, A, fitted with a stopcock on top, is inserted into the beaker so that the end of the tube is below the level of mercury in the beaker. Initially, the stopcock is open and the levels of mercury inside and outside the glass tube are the same. A vacuum pump is then attached to the top of the glass tube, and the air is pumped out. As this occurs, the mercury level rises in the tube until it reaches a height, h.

Despite further pumping, the mercury level rises no higher in the tube. When all air has been pumped out of the tube, the stopcock is closed, and the mercury maintains its position in the tube.

This behaviour is understood as follows. Initially, before pumping, the atmosphere exerts the same pressure on the surfaces of mercury inside and outside the tube. Since P_{out} = P_{in} = atmospheric pressure, the mercury levels inside and outside the tube are the same.

As air is pumped out of the tube, its pressure on the mercury inside the tube decreases, and mercury rises in the tube to a height, h, such that the force due to the mass of mercury in the tube just counterbalances the pressure of the atmosphere on the mercury outside the tube.

The height h is found to vary from day to day because the pressure of the atmosphere varies with weather conditions. However, its value is always in the neighborhood of 0.76 meters. For this reason, a unit of pressure called the atmosphere (atm) is defined as the pressure that will support a column of mercury 76 cm, or 760 mm, in height.

$$1 \text{ atm} = 760 \text{ mm Hg}$$

The pressure exerted by a column of mercury 1 mm in height is called a torr after Evangelista Torricelli, inventor of the barometer. Thus

$$1 \text{ atm} = 760 \text{ mm Hg} = 760 \text{ torr}$$

In the English and SI systems, the units of pressure are respectively the pound per square inch and the pascal (1 Pascal = 1 Newton per square meter, where a Newton is the force required to accelerate a 1 kg body at 1 m/s^2). The pressure in Pa exerted by a column of mercury 760 mm in height can be readily calculated and will provide a conversion factor between SI and traditional non-SI units. T

he force exerted by a mass, m, of mercury is mg, where g is the acceleration due to gravity. This force is distributed over the cross sectional area, A, of the tube. The mercury therefore exerts the pressure, P = mg/A on the surface of mercury at the bottom of the tube. The mass of mercury can be calculated from its density, r, and its volume, which is Ah. Substituting for m in the expression for pressure gives equation:

$$P = r^*(Ah)g/A = r^*gh$$

The pressure exerted by a column of mercury (or any liquid) of height h is independent of the cross sectional area of the column. The diameter of the barometer tube is therefore unimportant. Substituting the density of mercury (13.595 g/cm^3), the acceleration constant ($g = 9.80665\ m/s^2$), and h = 0.760 m in 5-1-1 gives $P = 1.0132 \times 10^5\ Nm^{-2}$ ($1\ Nm^{-2}$ = 1 Pascal). The following conversion relations are thus obtained:

$$1 \text{ atm} = 14.7 \text{ lb/in}^2 = 1.013 \times 10^5 \text{ Pa } (1 \text{ Pa} = 1\text{N/m}^2)$$

$$= 1.013 \text{ bar } (1 \text{ bar} = 10^5 \text{ Pa})$$

Although the bar is official, the atmosphere and the torr are still the most commonly used pressure units in chemistry. We will therefore use the atm and the torr.

Note that atmospheric pressure is 14.7/3.1 = 4.7 times greater than the pressure exerted by the feet of the man in Example. The magnitude of the pressure exerted by the atmosphere is equivalent to having that 150-lb man, holding a 550-lb weight, standing on your stomach.

The pressure of a gas confined in a container cannot be measured with a barometer. For this a variety of devices may be used, the simplest being the manometer. A manometer is a U-shaped glass tube containing mercury or some less toxic fluid. The gas container is attached to one side of the manometer, and the other side is either open to the atmosphere (an open-end manometer) or evacuated (a closed-end manometer).

The difference in the levels of fluid in the two arms of the manometer is a measure of the pressure of gas in the bulb through the relationship, $P_1 = P_2 + P$. Here P_1 is the pressure exerted by the gas on arm 1 of the manometer, P_2 is the pressure exerted on the arm of the manometer not connected to the gas container (arm 2), and P is the height of fluid in arm 2 minus the height in arm 1. P is positive if fluid is higher in arm 2 and negative if it is higher in arm 1.

For an open-end manometer, P_2 is atmospheric pressure; for a closed-end manometer, P_2 is zero. If the manometer fluid is mercury, the pressure is in torr and may be converted to atm using the conversion factor given above. If the manometer contains some other fluid, the height of fluid may be converted to an equivalent height of mercury using the densities of the fluid and mercury.

Example: An open-end manometer containing water as the working fluid is used to measure the pressure of gas in a bulb. The level of water in the side attached to the bulb is 8.65 cm higher than that in the side open to the atmosphere. Atmospheric pressure is 693 torr. What is the pressure of gas in the bulb?

Solution: Use the relation P = r*g*h. To find the height of mercury equivalent to 8.65 cm of water, we equate r*gh for water to r*gh for mercury. The gravitational acceleration cancels, giving

$$(1.00 \text{ g/cm}^3)(86.5 \text{ mm}) = (13.60 \text{ g/cm}^3)(h)h(Hg) = 6.4 \text{ mm}$$

The pressure of gas is this amount less than atmospheric pressure:

$$P_{gas} = 693 - 6.4 = 687 \text{ torr.}$$

IDEAL GAS LAW

The ideal gas law is a relationship between the pressure

(P), the volume (V), the amount (n), and the temperature (T) of a gas. The ideal gas law is called an equation of state, because it represents all possible states — combinations of P,V and T — in which the amount of gas, n, can be found.

$$PV = nRT$$

In this equation, P is the pressure in atm; V is the volume in L; n is the amount of gas in moles, and T is the Kelvin temperature. R is a proportionality constant, remarkably the same for all gases, which relates the 4 variables. Equation is of simple form, which belies the effort expended in developing it. Over 200 years elapsed between the first quantitative experiments on gases and the final enunciation of equation. The simple form of the equation hides a remarkable sophistication in understanding. We now examine what the equation says regarding gas behaviour.

VOLUME-PRESSURE RELATION

If we rearrange the equation to solve for the volume, the equation states that volume decreases as the pressure exerted on (also by) the gas increases. The inverse relation of pressure and volume is a fact of every day experience. For example, when you compress the air in the chamber of a tire pump, you feel the increase in pressure of the air within. In 1662, Robert Boyle developed the quantitative relationship between pressure and volume that we now know as Boyle's Law, which states that the volume occupied by a fixed amount of gas at a fixed temperature varies inversely with the pressure exerted on the gas. Boyle's Law was the first-discovered component of equation. Note that the ideal gas law predicts that the volume should fall to 0 as pressure gets very large. For real gases this does not happen.

At ordinary temperatures, most real gases will liquefy if pressure is made large enough. Once liquefication has occured, volume does not decrease much with increasing pressure. Real gases obey equation closely only when P is not too large. This is why equation is called the ideal gas law — it applies strictly only to an idealized gas, which does not liquefy but instead shrinks to zero volume when pressure is made very large.

VOLUME-TEMPERATURE RELATION

The ideal gas law states that the volume occupied by an amount of gas, n, varies directly with the absolute temperature of the gas. A particular sample of gas, then, should double in volume if heated from 300 to 600 K. This behaviour, too, is quite well known, at least qualitatively, in every day experience, as in the expansion (contraction) of a heated (cooled) balloon.

The relationship between gas volume and Celsius temperature was first established by Jacques Charles in 1800. He clearly established the linear relationship between the two quantities.

Similar experiments relating gas pressure and temperature were carried out at about the same time by Gay-Lussac, and similar proportionality was found. In both cases, the relationships obtained were not so simple as the ideal gas law, because the absolute scale of temperature was not then recognized.

The relation between volume and Celsius temperature is in equation,

$$V = a*t(^{o}C) + b$$

Plots of V versus t(°C) for gas samples containing different amounts of gas give different slopes and different intercepts on the V axis. However, they all give the same intercept on the t axis, -273.16 °C. This suggests that the zero of temperature be shifted 273.16 Celsius degrees to the left. The resulting scale of temperature is, of course, the Kelvin scale. Equation takes the much simpler form, V = a*T.

A profound implication of equation is that there exists a temperature at which the volume of gas should become zero. This temperature is called absolute zero. The Kelvin scale is set up so that 0 K corresponds to the temperature at which the volume of an ideal gas sample would become zero.

RELATION BETWEEN VOLUME AND AMOUNT OF GAS

The ideal gas law states that the volume occupied should increase in direct proportion with the amount (moles) of gas, when the gas is kept at the same pressure and temperature throughout. This is an extension of the hypothesis by Amadeo Avogadro in 1811 that equal volumes of different gases contain the same number of molecules.

Avogadro made this statement in explanation of many observations about the relative volumes of gaseous elements that react to form compounds.

It was in fact this hypothesis that finally allowed correct molecular formulas to be deduced and the atomic mass scale to be put on a firm basis.

VALUE OF GAS CONSTANT R

The gas constant R turns out to be an ubiquitous and extremely important quantity in physical science. We begin by obtaining a value for it based on the experimentally-measured volume of 1.0 mole of gas at 1.00 atm pressure and a temperature of 273.16 K.

It is found that under these conditions the gas occupies a volume of 22.414 L. Rearranging equation to solve for R, we can calculate its value:

$$R = PV/nT = (1.000 \text{ atm})(22.414 \text{ L})/(1.000 \text{ mole})$$
$$(273.16\text{K}) = 0.08206 \text{ L-atm/K-mole}$$

The numerical value of R depends upon the units in which pressure and volume are expressed. P is commonly expressed in atmospheres or torr (both non-SI) or Pascal (SI), and V is commonly expressed in mL. Values of R in common P-V unit combinations are presented in Table 5-1.

Table. In Common Pressure-Volume Unit Systems

Pressure Unit	Volume Unit	R
atm	L	0.08206 L-atm/K-mole
bar	L	0.08314 L-bar/K-mole
torr	mL	6.237×10^4mL-torr/K-mole
Pa	m^3	8.314 J/K-mole

From the Pascal-m^3 value, we see that R has units of energy/K-mole. It follows that the combined units L-bar, L-atm, and mL-torr are also energy units. We can express R in terms of any desired energy unit by using an appropriate conversion factor.

In using the ideal gas law, equation, the units of P, V, and R must be self-consistent, and the temperature must be expressed in Kelvins. A few examples of the use of the ideal gas law are now presented.

KINETIC MOLECULAR THEORY.

The ideal gas law was developed from macroscopic observations, with no knowledge of the behaviour of a gas at the molecular level. We now attempt to interpret the ideal gas law in terms of a reasonable molecular model (simple physical picture) of gas behaviour.

What are the individual molecules doing, and why does their behaviour manifest itself in the ideal gas law? We now enter the domain of theory — a mental interpretation of experimental results.

The model of gas behaviour currently accepted by scientists is the Kinetic Molecular model. The theory of gas behaviour built on this model is called the Kinetic Molecular Theory (KMT). This theory is one of the oldest and most resoundingly successful in science.

First, we assume that gases consist of very small molecules. Since a gas occupies the entire volume of its container, and since gases readily diffuse, the molecules must move about in space.

Consequently they have kinetic energy — energy of motion. Since gases are compressible, the molecules must be far apart.

Focussing on an individual molecule, it seems reasonable to assume that it is a small particle that moves about at random, occasionally colliding with a wall of the container and with other gas molecules.

The molecule sometimes gains and sometimes loses energy in these collisions, so that it frequently changes speed. Sometimes it moves rapidly, sometimes slowly. Thus our picture of the gas is dynamic rather than static. Figure is an attempt to portray this simple physical model of the gas phase.

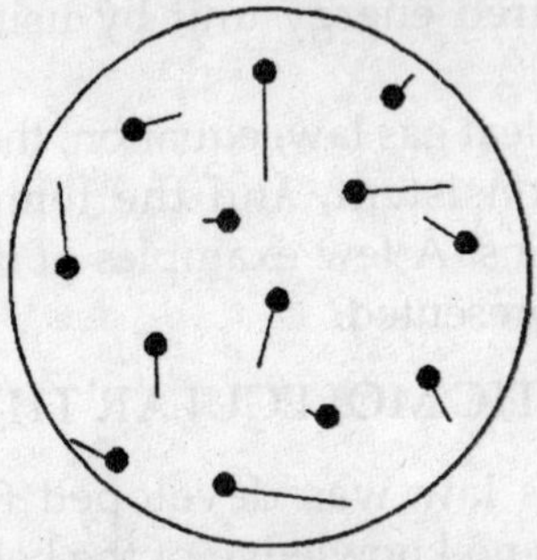

Fig. Dynamic Model of the Gas Phase

These statements about gas molecules are consistent with the properties of gases listed. None of them have been proven, but they seem reasonable and constitute the postulates of the Kinetic Molecular Theory:

- Gases consist of molecules that behave like tiny hard spheres;
- The molecules have no volume — they may be treated as points;
- There are no forces of attraction between molecules;
- Collisions between molecules are elastic (kinetic energy is conserved);
- The molecules are in ceaseless random motion, frequently colliding with the container walls and with each other. A collection of gas molecules is highly disordered.
- The average kinetic energy of the molecules is proportional to Kelvin temperature.

Our aim is to develop an expression for pressure. At the molecular level, pressure must result from collisions of gas molecules with the container walls. A molecule hitting the wall exerts a force on it.

The collection of all such forces on a unit area of wall during a given time interval constitutes the pressure. We should be able to calculate P from the force exerted by each collision, multiplied by the number of collisions per unit area of wall:

$$P = \text{Force/Area} = \text{Force/collision} \times \text{Collisions/area}$$

From Newton's second law, force is the rate of change of momentum.

Thus,

F = momentum change/time-collision × collisions/area

We move the time factor over to the second term to obtain equation.

F = momentum change/collision × collisions/time-area

To obtain the momentum change per collision, envision a molecule moving directly toward a wall with velocity v, as shown in Figure.

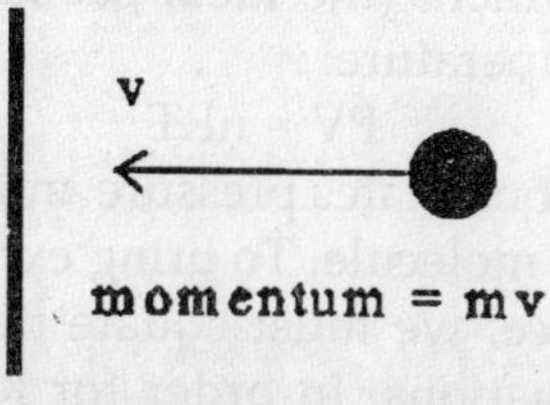

Fig. Collision of Gas Molecule with Container Wall

Since momentum is conserved in any collision, the change in momentum of the molecule is momentum change = momentum after collision - momentum before=

$$m(-v) - m(v) = -2mv$$

The change of momentum of the wall must therefore be 2mv to give a total change of zero. We now have the first term in equation:

momentum change/collision = 2mv

We take an intuitive approach to the second term. The number of collisions per unit time per unit area of wall should depend on 1) the number of molecules per unit volume in the container, N/V (the more there are, the more collisions there should be with the walls); and 2) the speed v at which a molecule moves (the faster the movement, the more collisions per unit time that should occur).

Equation is based on these ideas:

collisions/time-area = (N/V)(v)

If we assume for simplicity that the container is cubical, there are 6 walls over which these collisions must be spread.

The collisions with a particular wall are then,

$$\text{collisions/time-area} = Nv/6V$$

(Despite our restrictive assumptions about the shape of the container, equation turns out to be valid for a container of any shape!) We now multiply the expressions in equations to obtain the pressure:

$$P = (2mv)(Nv/6V) = Nmv^2/3V$$

Rearrangment gives equation,

$$PV = Nmv^2/3$$

This is as far as our molecular model takes us.

We are now at the interface between theory and experiment. Experiment (the ideal gas law) relates pressure and volume to temperature:

$$PV = nRT$$

Theory, equation, relates pressure and volume to the mass and speed of a gas molecule. To bring experiment and theory into correspondence, we must equate the right sides of the preceding two equations. In order for KMT to successfully explain the ideal gas law, it is necessary that,

$$nRT = Nmv^2/3$$

Equation can be simplified by replacing n with N/N_o (N_o is Avogadro's Number); $mv^2/2$ by KE (the kinetic energy of an average gas molecule); and solving for KE.

The result is,

$$KE_{molecule} = 3RT/2N_o$$

We refine this equation by making two further realizations. First, since R and N_o are both constants of nature, so must their ratio be. It is symbolized k, and is called Boltzmann's constant. Its value is 1.381×10^{-23} J/K-particle. Second, we have recognized that the kinetic energy of a gas molecule changes frequently as it undergoes collisions. However, its average speed over time is constant. The refined equation is among the most important equations in science:

$$\text{Average } KE_{molecule} = 3kT/2$$

The average kinetic energy of a gas molecule, no matter what its chemical identity, depends only on the Kelvin temperature. This gives a deep insight into the concept of temperature: it is a measure of the average kinetic energies of the molecules of a substance, hence a measure of their average speed. As T is increased, molecules move faster; as it is lowered, they move slower. The temperature at which molecular motion ceases is absolute zero. Temperatures lower than absolute zero are impossible because a molecule may not have negative kinetic energy.

Equation is simple and profound. Is there an experiment that can be done to test its validity? There is in fact a simple experiment that verifies equation in a rearranged form. If we replace the average kinetic energy of a molecule with the expression $m(v^2)_{avg}/2$, where $(v^2)_{avg}$ is the average of the square of the speed, and solve the resulting expression for $(v^2)_{avg}$, we obtain equation.

$$(v^2)_{avg} = 3kT/m$$

The square root of $(v^2)_{avg}$ is called the root mean square speed, and is for our purposes approximately equal to the average molecular speed. Making this equality gives equation.

$$v_{avg} = (3kT/m)^{1/2}$$

Multiplying both numerator and denominator of the argument on the right side of this equation by Avogadro's number N_o gives the useful variant in equation.

$$v_{avg} = (3RT/MM)^{1/2}$$

The implication of equations is that a heavy gas molecule moves more slowly than a light one, in a quantifiable way. The ratio of the speeds of the light (L) and heavy (H) molecules should be the square root of the inverse ratio of their molar masses:

$$(v_H/v_L)_{avg} = (MM_L/MM_H)^{1/2}$$

In 1846, Thomas Graham measured the rates of diffusion of various gases. Diffusion is the process by which gases disperse in space via random molecular motion. The results of Graham's experiments are summarized in equation rate diffusion of gas A/rate of diffusion of gas B =(density of gas B/ density of gas A)$^{1/2}$

ROLE OF GAS PHASE CHEMICAL REACTIONS

The first decade of the 19th century must have been both

exhilarating and frustrating for chemists. On the one hand, Dalton's atomic theory (1803) rationalized many known facts about chemical elements, compounds, and reactions. In particular, the theory provided an explanation for the laws of conservation of mass, definite proportions, and multiple proportions, and gave a rational basis for the idea of reproducible combining masses of the elements in compound formation (e.g., the combining masses of oxygen and hydrogen in forming water are in the ratio 7.94 to 1). On the other hand, attempts to develop a scale of atomic masses were thwarted by the ignorance of chemical formulas. Chemists, led by Dalton, assumed that the formula for water was HO, and that water was formed from hydrogen and oxygen by equation:

$$H + O \rightarrow HO$$

This was the simplest assumption to make in the absence of definite knowledge of the formulas of elemental hydrogen, elemental oxygen, and water.

In 1809, Joseph Gay-Lussac conducted a series of experiments with gases that ultimately enabled the development of the atomic mass scale. Specifically, for reactions in which gases react to form other gases, he studied the relationships among the volumes of gaseous reactants consumed and the volumes of gaseous products formed. His result is of tremendous importance. Gay-Lussac discovered that, as long as the experiments were conducted at a constant temperature and pressure, the volumes of reactants used and products produced always gave whole-number ratios. One of many reactions that he studied was that between hydrogen and oxygen to form water vapour. He found consistently and reproducibly that for every one volume of oxygen used, two volumes of hydrogen were required, and two volumes of water vapour were produced:

(5-4-2): 2 vol hydrogen + 1 vol oxygen $\rightarrow$ 2 vol water vapour (5-4-2)

Gay-Lussac was not certain what to make of these results; however, Amadeo Avogadro was. He made the (to him) reasonable proposal that Gay-Lussac's results implied that equal volumes of gases, at the same temperature and pressure, contain equal numbers of molecules. In modern terms, we say that the volume occupied by a gas is proportional to the number of moles of gas present.

Operating from this assumption, he concluded and stated that Dalton's view of the hydrogen/oxygen reaction, expressed in 5-4-1, was incorrect, because it was not consistent with the two-to-one-to-two hydrogen/oxygen/water volume ratios that Gay-Lussac observed. He then went on to propose that hydrogen and oxygen occur as diatomic molecules, and that water in fact contains two atoms of hydrogen per atom of oxygen, so that the water formation reaction becomes:

$$(5\text{-}4\text{-}3): 2H_2(g) + O_2(g) = 2H_2O(g)$$

This is the simplest proposal consistent with the combining volumes observed; as we know today, it is entirely correct.

GAS PRODUCTION

Three broad classifications of chemical reactions: electron transfer processes, proton transfer processes, and double displacement processes. We defined a double displacement reaction as one in which the positive and negative portions of the reactant molecules interchange to form products. It is appropriate on gases to focus briefly on double displacement reactions that produce a gas as one of the products. An important series of examples of such reactions involves metal carbonate compounds (called carbonate salts) and acids as reactants. The reaction of a metal carbonate with an acid is shown generically in equation.

$$MCO_3(\text{s or aq}) + HX(aq) = MX_2(aq) + H_2CO_3(aq)$$

This reaction does not have a gas as one of its products. However, in a subsequent step, carbonic acid, H_2CO_3, decomposes to water and carbon dioxide, a gas that escapes in part from the aqueous solution in the form of effervescence:

$$H_2CO_3(aq) = H_2O + CO_2(g)$$

A specific example of the acid-carbonate reaction. Here the net reaction, obtained as the sum of equations, is shown.

$$Na_2CO_3 + 2\ HCl = 2NaCl + H_2O + CO_2$$

Acid-carbonate reactions are very important in at least two contexts. First, metal carbonates that have low solubility in water make up a substantial fraction of the earth's crust (limestone is an example).

Deposits of carbonate rocks in contact with slightly acidic ground waters undergo reactions. In so doing, they serve to maintain the acidity level of the water at a fairly constant level, supporting aquatic plant and animal life.

Second, reactions are important in the context of acid rain, which results when nitric acid, HNO_3, and sulfuric acid, H_2SO_4, are produced in the atmosphere from SO_2, produced in coal combustion, and NO_2, a byproduct of the internal combustion engine. Limestone (primarily calcium carbonate) is widely used in construction, and has been used for centuries as a medium for sculptors.

Repeated and prolonged contact of building edifices and priceless works of art with acid rain results in their slow destruction as calcium carbonate reacts with nitric and sulfuric acids according to equations. Building edifices can be replaced; works of art cannot.

Finally, dietary calcium tablets are made primarily of calcium carbonate. Contact of a stomach tablet with stomach acid (hydrochloric acid) causes the tablet to dissolve via reaction, after which the calcium can be utilized by the body.

$$CaCO_3(s) + HCl = CaCl_2(aq) + H_2O + CO_2(g)$$

Metal sulfites (for example, Na_2SO_3) react with acids in analogous fashion to form aqueous sulfurous acid, H_2SO_3. This subsequently decomposes to water and sulfur dioxide:

$$Na_2SO_3(aq) + HNO_3(aq) = NaNO_3(aq) + H_2SO_3(aq)$$

$$H_2SO_3(aq) = H_2O + SO_2(g)$$

Chapter 2

Properties of Substances

The vast majority of known substances are solids under ordinary conditions of pressure and temperature. An understanding of the structures of crystalline solids at the atomic/molecular level. Be forewarned that a study of such structures requires a sound basis in geometry and an ability to visualize in 3-dimensions.

Further, it is impossible to obtain an adequate understanding of the important structural ideas from text and 2-dimensional figures. Work with 3-dimensional models is essential. Understand it as best you can, then refine your understanding with models in the laboratory. A solid base for the understanding of two major classes of materials: metals, in which all atoms in the structure are the same; and ionic compounds, composed of cations of one substance and anions of another.

MACROSCOPIC PROPERTIES OF SOLIDS

Solids have a number of observable properties in common:

- They are rigid, maintaining fixed shape and volume independent of container.
- The molar volumes of solids are small.
- Solids are incompressible.
- Solids do not flow, and diffusion in solids is extremely slow.
- Many solids occur as crystals — well-defined shapes with structures suggesting an orderly internal arrangement of atoms, molecules, or ions.

Figure shows a 2-dimensional representation of a solid.

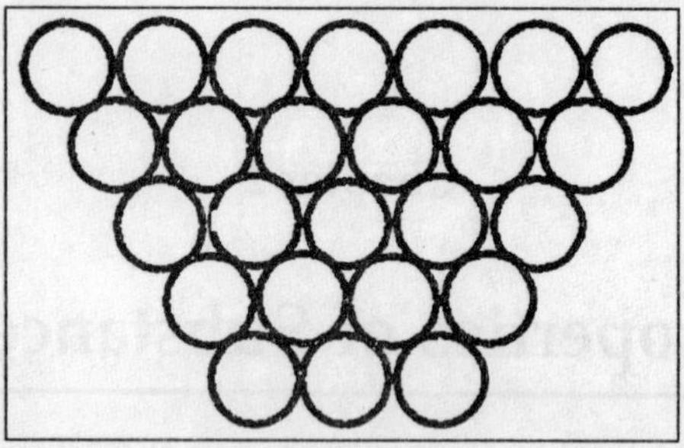

Fig. 2-dimensional Representation of a Solid

Note first how close together the particles are compared with the gas phase. This crowding allows very little free volume to the molecules, which strongly resist further compression. The solid is incompressible. It is also clear why diffusion is slow in solids, because each particle is locked in place by the surrounding ones. Only at the surface of the solid, where a particle has neighbors on only 3 sides, can diffusion occur.

The particles are in constant motion, just as in the gas phase. A molecule is constantly jiggling in place, bumping into neighbors that hold it in a cage. Just as for gases, the kinetic energies of the particles (atoms, molecules, or ions) of the solid are distributed according to the Maxwell-Boltzmann distribution, with the kinetic energy of an average particle having the value 3kT/2. The major difference between the motion in gas and solid is that in the solid, a particle is hemmed in and undergoes many more collisions per unit time than in the gas. The movement of a particle in the solid is much like a vibration, because the molecule stays in the same location.

If the temperature of the solid is increased, the particles vibrate faster, but are still not free to move about because intermolecular forces from the neighbors hold them in place. Finally, the arrangement of particles is very orderly, not at all chaotic like the gas. Each particle is surrounded by a perfect hexagon of other particles; and particles are arranged in rows that remain straight over long distances. It is easy to imagine that this orderly arrangement at the molecular level is manifested in beautiful crystalline shapes at the macroscopic level.

LATTICE STRUCTURE

A lattice is a regular, repeating pattern of points in space. Wallpaper patterns are examples of 2-dimensional lattices. The lattice is more fundamental than the particular wallpaper pattern shown, because wallpaper patterns using many different design units could be generated from the same lattice. The key feature of the lattice is the arrangement of points. A small collection of lattice points that contains all of the spatial information in the lattice pattern is called a unit cell. The unit cell represents the lattice, because the whole lattice can be reproduced from the unit cell by moving (translating) the unit cell in directions parallel to its sides by distances equal to the lengths of its sides.

Figure shows a second small collection of points that might be considered as a unit cell. Movement of this collection in directions parallel to its sides does indeed reproduce the entire lattice.

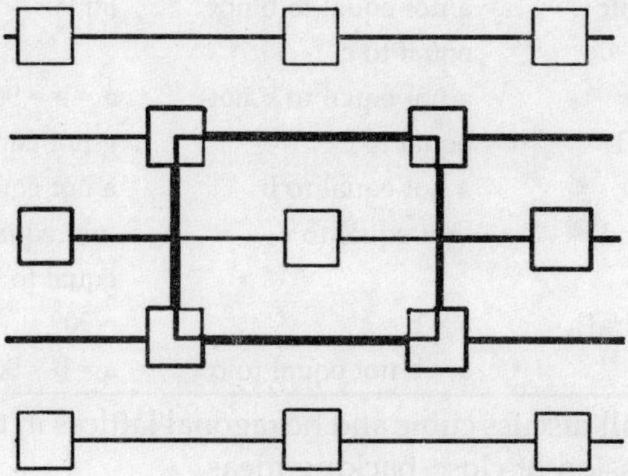

However, the unit cell in Figure above is simpler because it involves fewer points — one point at each of its 4 corners. A unit cell having lattice points at its corners only is called a primitive unit cell. The unit cell in Figure above is a non-primitive unit cell because it has an additional point at its centre. If it is possible to represent a lattice with a primitive unit cell, we will do so.

Crystalline solids are represented by 3-dimensional lattices, one example of which is the simple cubic lattice. This is a primitive unit cell because it has points only at the corners. Movement of the unit cell by the distance, d, in directions parallel to its 3 dimensions reproduces the entire simple cubic lattice. The simple cubic lattice is analogous to an orderly arrangement of building blocks stacked face to face in 3-dimensions.

It was shown in the 1800s by Bravais that there are only 7 shapes (called crystal types) that a unit cell can have and only 14 possible distributions of points (lattice types) within these 7 shapes. The 7 shapes, characterized by the relative magnitudes of the repeat distances in three directions and the angles between these directions, are summarized in Table.

Table. The Bravais Crystal Types

System	Edge Lengths	Angles
Cubic	a = b = c	all 90°
Tetragonal	a = b not equal to c	all 90°
Orthorhombic	a not equal to b not equal to c	all 90°
Monoclinic (once-angled)	a not equal to b not equal to c	a = b = 90°, g not equal to 90°
triclinic (thrice-angled)	a not equal to b not equal to c	a not equal to b not equal to g not equal to 90°
Rhombohedral	a = b = c	= 90°
Hexagonal	a = b not equal to c	a = b = 90°; g = 120°

We will discuss cubic and hexagonal lattices in the context of a discussion of close-packing ideas.

CLOSE PACKED STRUCTURE

What is the most efficient way of arranging spheres in 3-dimensional space? We take "efficient" to mean economical in terms of space. We begin by building one layer of spheres that are arranged as closely to one another as possible. Note that this arrangement is identical to that pictured earlier. Spheres are lined up in straight rows, with the spheres in one row fitting snugly into the indentations in adjacent rows. Each sphere is surrounded by a perfect hexagon of spheres, all touching it. There are numerous gaps, or holes, in the layer, of triangular shape. For descriptive purposes here, we label these holes "u" if the triangle has a single vertex pointing up, or "d" if the triangle has vertex down.

We now add a single sphere of a second layer on top of the first. The added sphere rests in a low point of the first layer, corresponding with a triangular hole, in contact with 3 spheres of the first layer. The 4 touching spheres lie at the vertices of a regular tetrahedron. There remains some unfilled space at the centre of this 4-sphere group that is called a tetrahedral hole. Each time we add a sphere to the second layer, we create a tetrahedral hole between the layers.

When additional spheres are added to the second layer, it turns out that they must all go in either "u" holes or "d" holes, but not both. Thus only half of the triangular depressions in the first layer can be used for placement of 2nd-layer spheres.

Triangular holes in the 2nd layer lie directly over the unused holes of the first layer, so that channels pass right through both layers. The point at the centre of such a channel is surrounded by 6 spheres, arranged as 2 equilateral triangles, one triangle in the 1st and one in the 2nd layer. These 6 spheres form a regular octahedron, a geometrical solid with 6 vertices and 8 equilateral triangular faces.

The space surrounded by the 6 spheres is called an octahedral hole. An octahedral hole is larger than a tetrahedral hole because 6 spheres cannot be crowded as closely together as 4 spheres.There are 2 ways in which a third layer of spheres can be added to the existing 2 layers in close-packed fashion. In the first way, spheres are place directly over spheres of the first layer. The third and first layers then have spheres in the same locations.

The 4th layer then goes over the second, the 5th over the 3rd, and so on, so a pattern that repeats every other layer is obtained. This is characterized as a 1212... pattern. The unit cell is hexagonal in shape; the 1212... close-packing arrangement is therefore called hexagonal close packing (hcp).

Note that the coordination number (the number of nearest neighbors) of any sphere in hcp is 12.

The second way of adding a third layer is to place spheres directly over the octahedral holes between the first two layers. In this case, the third-layer spheres do not lie over spheres in layer 1. This is still a close-packing arrangement because the coordination number of any sphere is 12, just as in hcp. The 4th layer of spheres is then placed directly over the spheres in the first layer. This type of close packing is designated 123123... to indicate that the pattern repeats after 3 layers. The unit cell for 123... close packing. Although it may not be obvious from the figure, this unit cell is a cube (this is a situation in which a 2-dimensional drawing does not adequately convey what is true). This type of close packing is therefore called cubic close packing (ccp).

Hcp and ccp are the two ways of close packing spheres in 3 dimensions.

They have several features in common:

- Spheres occupy 74% of the available space;
- Each sphere has 12 nearest neighbors that touch it (coordination number);
- The number of octahedral holes in the lattice is exactly the same as the number of spheres in the lattice, and there are twice as many tetrahedral holes as spheres! This is not obvious and will be addressed again later.

There are a number of substances that crystallize in one or the other of these two packing arrangements. Beryllium, magnesium, and zinc metal all form hexagonal close packed lattices. Nickel, copper, gold, and aluminum form ccp lattices. Cubic close packing is particularly common. We will take a closer look at cubic lattices now.

The three Bravais cubic lattices. The simple cubic (sc) lattice is primitive, with points (spheres or atoms) only at the vertices of the cube. Metallic polonium and non-metallic dioxygen (O_2) crystallize with the simple cubic lattice. The coordination number of an atom in the simple cubic lattice is 6. To determine this, pick one atom in the unit cell; find an atom that appears to be closest to the atom you picked; then count the number of other atoms that are at this same distance from the atom you picked.

In counting coordination numbers, you may have to go outside of the single unit cell. In the simple cubic lattice, identical unit cells surround the pictured one on all sides. Some of the nearest neighbors of your picked atom lie in these neighboring unit cells.

Finally, the simple cubic unit cell contains only 1 atom, despite the fact that you see 8 atoms drawn in the figure. We say that the simple cube has 1 atom per unit cell (1 atom/uc). The procedure for counting atoms in unit cells is described in the paragraph below.

The body-centered cubic (bcc) lattice is non-primitive, with points at the corners and the centre of the unit cell (the centre of a cube is called the cube body centre). Metallic iron and barium crystallize in the bcc lattice.

The coordination number is 8 for this lattice (count the nearest neighbors of the body-centre atom), and the body-centered cube has 2 atoms per unit cell. Finally, the face-centered cubic (fcc) lattice is also non-primitive with points at each corner and in the centre of each of the 6 faces of the cube.

Metallic nickel and copper adopt this structure. The coordination number is 12 for this lattice, which means that it must be close packed! The face-centered cubic lattice and the cubic close packed lattice are identical. It is very difficult to see this in two-dimensional drawings.

To see it, you must imagine looking along a body diagonal of the face-centered cube. A body diagonal runs on a straight line from one corner, through the body centre of the cube, to the diagonally opposite corner.

As you look along the body diagonal, you will see the pattern of spheres: a single sphere; a triangle consisting of 6 spheres; a second triangle of 6 spheres, but rotated 60° with respect to the first one; and finally another single sphere immediately below the first one. The cube is admittedly difficult to visualize from the drawings. The face-centered cubic lattice has 4 atoms per unit cell.

We now address the problem of determining the number of atoms per unit cell. This is quite simple once it is realized that atoms are generally shared among several unit cells and therefore must be partitioned accordingly.

An atom at a corner (vertex) of a cube is shared among 8 unit cells. To see this, imagine the simple cube in Figure to be a building block, and focus on the upper right front corner of the building block. Imagine that this corner is labelled "a".

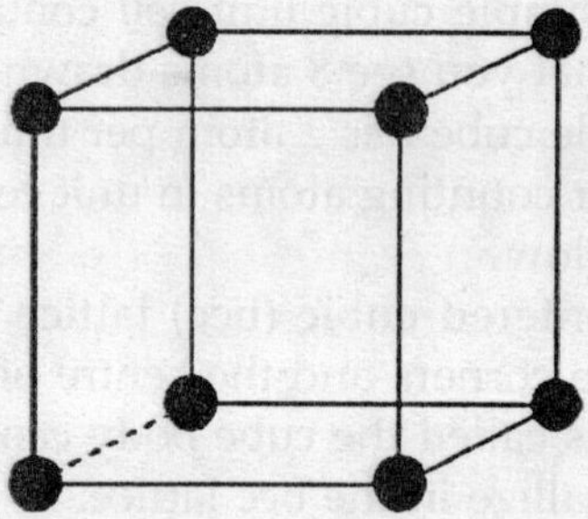

In your mind, place identical blocks on top, to the right, and in front of the block in the figure. Then place additional blocks in the remaining places adjoining these blocks to generate the 8-block. Point "a" is now at the centre of this collection, and is simultaneously a corner for each of the 8 blocks; 8 blocks share a vertex. This point cannot be said to belong entirely to any particular block. Instead, each block gets 1/8 of it. The counting rule for a corner atom, then, is that only 1/8 of it is assigned to a single unit cell. Similar thinking leads to the conclusion that a face-centre atom is shared between 2 unit cells, so 1/2 of it is assigned to a particular unit cell; and a body-centre atom is not shared at all, so it belongs entirely to the unit cell that it is found in. The atom count for a body-centered unit cell then goes like this:

atoms/uc = 8 corner atoms × 1/8 + 1 body-centre atom x 1

We will have occasion also to partition points located along edges of the cube. Edges are the lines connecting adjacent vertices. Since an edge is common to 4 cubes arranged in a square, 1/4 of any point along an edge is assigned to a particular cube. The point-counting rules are summarized in Table below.

Table. Atom Counting Rules for Cubic Unit Cells

Atom location	Fraction Assigned to uc
Corner	1/8
Edge	1/4
Face	1/2
Body-centre	1

If, as we have said, the face-centered cubic lattice is in fact cubic close packed, then there should be tetrahedral and octahedral holes in the fcc lattice. Further, we should find 4 octahedral and 8 tetrahedral holes in a fcc unit cell based on the the last-listed common feature of close-packed lattices above. To locate these holes it is convenient to place our cube in a coordinate system. The origin of coordinates is located at the lower left front corner of the cube, with the x-axis along the bottom front edge, the y-axis along the left vertical edge, and the z axis along the left bottom edge.

We specify distances along the 3 axes in terms of fractions of the cube edge length. The set of coordinates (1,0,0) means to proceed 1 edge length along x. These are the coordinates of the lower right front vertex of the cube (labelled 1 in the figure). Similarly, (1/2,1/2,1/2) specifies the body centre of the cube, labelled 2 in the figure.

Now examine the face-centered cube. We find the most obvious octahedral hole at the body centre. This point is surrounded by 6 atoms all at the same distance from it. These are the atoms at the face centers of the cube. Similar octahedral holes are located at the cube edge centers.

It is surrounded at equal distance by the atoms at (1/2,1/2,1), (1,0,1), (1,1,1), and (1,1/2,1/2). The remaining 2 atoms are not part of the unit cell shown, but can still be located with coordinate sets (1 1/2,1/2,1) and (1,1/2,–1 1/2). (Translating these coordinates is good practice in the use of the unit cell coordinate system.) The cube has 12 edges, so there are 12 holes of this type.

The counting rules tell us that only 1/4 of each of these edge-centered octahedral holes belongs to the unit cell of interest. The total number of octahedral holes in the unit cell is then 1(body centre) + 12(1/4) = 4. The number of octahedral holes is the same as the number of atoms in the unit cell!

A tetrahedral hole in the fcc unit cell. It is defined by the four shaded atoms, and has coordinates (1/4,1/4,1/4). There are 7 other tetrahedral holes in the unit cell, all with coordinate combinations involving 1/4 and 3/4, and all defined by 1 corner and 3 face-centre atoms. Try to locate them in the figure, and write down their coordinates. Each tetrahedral hole lies entirely within the cube body, so none of them are shared with other unit cells. The number of tetrahedral holes per fcc unit cell is therefore 8, twice the number of atoms in the unit cell.

Octahedral and tetrahedral holes are important in the discussion of ionic compounds.

IONIC COMPOUNDS

The ideas discussed thus far apply to crystals in which the units — spheres or atoms — are all identical. Crystals of pure metals, with close-packed metal atoms, and even crystals of molecular solids (for example, methane, CH_4, in which there is a methane molecule at each lattice point) are examples of real situations in which these ideas apply. Ionic compounds are different because there are two distinctly different units — the cation and the anion — that must be accommodated in the lattice. The resulting lattices are often called compound lattices because they contain two different components. The cation and the anion differ in charge, of course, and in chemical identity (usually). They also differ in size. Generally speaking, cations are smaller than anions, for easily-understood reasons. Cations contain an excess of protons over electrons. The Coulomb pull of the excess positive charge causes the electron cloud to contract, so that it is smaller than in the corresponding atom (e.g., Na^+ is smaller than Na).

Anions contain an excess of electrons over protons. The repulsion between the electrons exceeds that in the corresponding neutral atom, causing an expansion of the electron cloud (e.g., Cl^- is larger than Cl). Except for a few rare cases, the cation in an ionic compound is smaller than the anion. In forming a lattice, the space requirements are more important for the larger anion, which tends to dictate which lattice is adopted. The (smaller) cations then fit into the holes, or interstices, of the anion lattice. Many common ionic compounds have lattices that consist of cubic close packed anions, with cations in tetrahedral and/or octahedral holes. Stoichiometries of up to 3 cations per anion can be accommodated in this framework. We now discuss some common ionic lattices.

The Sodium Chloride (Rock Salt) Lattice. It can be described as a cubic close packed (face-centered cubic) anion lattice, with Na^+ ions in the octahedral holes. The coordination number (number of nearest neighbors of opposite charge) of each Na^+ ion is 6, and of each Cl- ion is 6. The cation and anion must have equal coordination numbers because the stoichiometry of the compound is 1:1 (1 cation per anion). The unit cell contains 4 formula units of NaCl (4 Cl- ions, which define the fcc unit cell; 4 Na^+ ions because there are 4 octahedral holes per unit cell). Many ionic compounds exhibit the NaCl lattice, including most of the group 1 halides, most of the group 2 oxides, and a number of d-metal oxides.

The Zinc Blende Lattice. Zinc blende is the common name for one crystal modification of the common mineral, ZnS. The other form is known as wurtzite. Wurtzite consists of an hcp array of S^{2-} ions with Zn^{2+} ions in one-half of the tetrahedral holes; it will not concern us further here, because we are restricting our attention to cubic lattices. Zinc blende consists of a ccp array of S^{2-}, with Zn^{2+} in half of the tetrahedral holes.

Since there are 8 tetrahedral holes in the sulfide ion lattice, but only 4 S^{2-} ions, only half of the holes are used to accommodate 4 Zn^{2+} ions. This is consistent with the 1:1 stoichiometry of the compound.

The zinc ions occupy tetrahedral holes so that they can be as far apart from one another as possible to minimize repulsions between positive ions. If the first Zn^{2+} ion occupies the hole at (1/4,3/4,1/4), the other three will be in holes at (3/4,1/4,1/4), (1/4,1/4,3/4), and (3/4,3/4,3/4). You should locate these holes and verify that the Zn^{2+} ions describe a tetrahedron. The cation and anion coordination numbers are both 4 in zincblende. As for NaCl, there are 4 formula units per unit cell. Other ionic compounds that adopt the zinc blende lattice include BeS, CdS, HgS, and AgI.

The Calcium Fluoride (fluorite) Lattice. This compound has formula CaF_2, and exhibits. The Ca^{2+} ion is virtually the same size as the F- ion, one of those rare situations referred to above, and forms a face-centered cubic lattice. The F- ions fill all of the tetrahedral holes in the cation lattice. Since there are 8 such holes per 4 Ca^{2+} ions, the stoichiometry is nicely accommodated.

The coordination number of the Ca^{2+} ion is 8, and that of the F- ion is 4. (Note that the number of cations per formula unit multiplied by the cation coordination number is equal to the product of the number of anions per formula unit and the anion coordination number.) There are again 4 formula units per unit cell. The fluorite structure is very common for ionic compounds of 1:2 (or 2:1) stoichiometry.

The Cryolite Lattice. Cryolite is the common name for the mineral, Na_3AlF_6. Though rare, this substance is of extreme commercial value, because it is used in the process of extraction of aluminum from bauxite ore. Cryolite consists of Na^+ cations and AlF_6^{3-} anions. It adopts a lattice in which the anions are cubic close packed (fcc) and the cations occupy all of the octahedral and tetrahedral holes.

The Diamond Lattice. Diamond is one crystal modification of the element carbon. It is, of course, not ionic, nor even a compound. It is dealt with here because it, too, is based on a fcc unit cell. The diamond lattice is just like the zinc blende lattice, with carbon atoms in place of both Zn^{2+} and S^{2-} ions. We therefore describe it as a fcc array of carbon atoms with carbon atoms in one-half the tetrahedral holes. The coordination number of each carbon atom in diamond is 4.

The Cesium Chloride Lattice. The cesium chloride lattice. It is a more open lattice than those considered thus far. This lattice consists of a simple cubic array of chloride ions, with Cs^+ ions in cubic holes. A cubic hole is the space at the body centre of a simple cubic unit cell. There is 1 formula unit of CsCl per unit cell, and the coordination number of both cation and anion is 8. Ionic compounds that adopt the CsCl lattice include CsBr, CsI, the halides of thallium(I), and NH_4Cl.

Radius Ratio Rules. Compounds of 1:1 stoichiometry commonly adopt one of three lattices discussed above: the zinc blende lattice, the rock salt lattice, or the cesium chloride lattice. Of interest now is whether or not we can predict, or at least understand, why a particular compound favors one lattice over the others. A number of factors are important in choice of lattice, but one of these dominates in importance: the relative sizes of cation and anion.

Because we view the cation as occupying holes in the anion lattice, we begin by calculating the sizes of the various hole types in relation to the size of the anion involved. The most straightforward calculation is for the octahedral hole. The figure shows one face of the fcc unit cell, with the anions just touching along a face diagonal of the cube.

There is an octahedral hole at the centre of each of the 4 cube edges shown; we focus on the right edge. The size of the octahedral hole will be taken as the distance from the edge centre to, say, the top right anion, along the cube edge. (The distances from the hole to the bottom right anion and to the face-centered anion are the same.) We can express the right edge length as a sum of anion and hole radii as in equation:$a = 2r^- + 2r^+$

Here r^- is the anion radius, r^+ the hole radius. (The + designation is used in anticipation of putting a cation in the hole.) Similarly, we express the length of the face diagonal (fd) in terms of anion radii.

This is easy, since the anions touch along this line:

$$fd = 4r^-$$

Finally, we relate the fd and edge length via the Pythagorean.

Theorem:

$$(fd)^2 = 2a^2$$

Take the square root of both sides of 7-4-3:

$$fd = a(2)^{1/2}$$

Substitute 7-4-4 in 7-4-2 and rearrange for a:

$$a = 2(2)^{1/2}\, r^-$$

Substitute for a in 7-4-1 and solve for r^+:

$$r^+ = ((2)^{1/2} - 1)\, r^- = 0.414\, r^-$$

The octahedral hole is somewhat less than half as large as the anions that create the hole. By similar geometric methods, equations for tetrahedral and cubic holes respectively, can be developed:

$$r^+ = 0.224\, r^- \quad r^+ = 0.732\, r^-$$

You should try to obtain these expressions, making use of right triangles and Pythagorean relationships. (In addition to equation, which relates the face diagonal and edge length of a cube, equation relates the cube body diagonal (bd) to the edge length:

$$bd = (3)^{1/2}$$

A tetrahedral hole is about 1/2 the size of an octahedral hole and about 1/3 the size of a cubic hole. Only cations that are relatively small compared with the anion can be accommodated in tetrahedral holes. As cation size increases relative to anion size, the cation will prefer a larger hole, with a larger coordination number of anions. Equations are the basis for the following radius-ratio rules.

- If the ratio of the cation and anion radii (r^+/r^-) lies in the range between 0.224 and 0.414, the cation prefers to occupy a tetrahedral hole.
- If r^+/r^- lies in the range between 0.414 and 0.732, the cation prefers an octahedral hole.
- If r^+/r^- exceeds 0.732, the cation prefers a cubic hole.

These rules make good sense. If $r^+/r^- < 0.224$, the cation is too small for the tetrahedral hole. The anions will be touching, but will not be touching cations. Since it is the cation-anion attractions that stabilize the lattice, whereas anion-anion or cation-cation repulsions tend to destabilize it, this is not a favorable situation.

When $r^+/r^- = 0.224$, the cation just fits the tetrahedral hole. Although anions are still touching, they now touch cations as well. This is a stable arrangement. As r^+/r^- becomes > 0.224, the cation is too big for the tetrahedral hole and starts to push the anions apart.

The cation-anion attractions are maintained, because cations and anions still touch, but now anion-anion repulsions are decreased because the anions no longer touch. This is an even better arrangement than the perfect fit.

This situation continues to improve until r^+/r^- becomes equal to 0.414. At this value, the cation prefers to occupy an octahedral hole because it can increase its coordination number without increasing anion-anion repulsion to any significant extent. The number of cation-anion attractive interactions increases from 4 to 6, giving a more stable lattice.

As r^+/r^- increases between 0.414 and 0.732, cations in octahedral holes maintain their contact with 6 anions while pushing the anions further and further apart, decreasing repulsions between them.

Throughout this range of r^+/r^-, occupation of an octahedral hole becomes increasingly favorable until the radius ratio becomes 0.732. The cation can now increase its coordination number to 8 anions, again increasing the number of attractive cation-anion interactions. The radius ratio rules are summarized in Table.

Table. Radius Ratio Rules

Range of r^+/r^-	Lattice Type	Example
0.224-0.414	ZnS	ZnS
0.414-0.732	NaCl	KCl
> 0.732	CsCl	CsCl

Ion radii and the radius ratio rules can be used to predict the probable choice of lattice for an ionic compound.

X-RAY CRYSTALLOGRAPHY.

In 1912, von Laue predicted that crystals could serve as diffraction gratings if the wavelength of radiation used were approximately the same as the distance between planes of atoms in the crystals. This spacing was known to be on the order of 1×10^{-10} m, corresponding to wavelengths in the x-ray region of the electromagnetic spectrum. Subsequently, WH Bragg discovered that when X-rays of appropriate wavelength are directed onto the smooth face of a crystal, the X-rays are diffracted and a pattern characteristic of the lattice structure of the crystal is obtained.

This observation is the foundation for the modern technique of X-ray Crystallography, the most powerful method that we have for determining molecular structure. Like the types of spectroscopy, x-ray crystallography is based on the interaction of light with matter. Unlike spectroscopy, however, the light is not absorbed by the matter in this case. Instead, it is scattered from the atoms of the crystal in a very definite way. Analysis of the scattering enables the arrangements of atoms within the crystal to be deduced. The analysis of x-ray diffraction patterns is complex, even for materials of simple structure. However, thanks to the efforts of many crystallographers over many years, it is today done quite quickly and routinely.

Once a suitable crystal of a substance is grown, the modern crystallographer is able to mount it in a fully automated x-ray diffractometer, the instrument that is used for the x-ray diffraction study. The diffractometer automatically obtains all of the necessary data for the determination of the lattice structure. The data are then fed to one of a number of computer programs that have been developed and refined for the analysis of such data, and the structure is "solved."

The output of the programme reveals not only the lattice structure, but also the locations of all of the atoms of a molecule of the substance. Structures of substances ranging from simple salts or molecular substances to huge complex proteins and enzymes can be determined using this approach.

LATTICE IMPERFECTIONS

Perfect crystals, in which every atom or molecule is in place, are an idealization, like the ideal gas. Real crystals contain fairly large numbers of imperfections, called defects. Three major types of defect are called point defects, line defects, and plane defects. Common point defects in simple lattices (in which all lattice points are identical) are vacancies (in which an atom is missing); atoms out of place, occupying what is normally a hole in the lattice; substitutional impurities, in which foreign atoms occupy occasional sites; and interstitial impurities, in which foreign atoms occupy holes in the lattice. Ionic (compound) lattices exhibit Schottky defects, in which a cation or anion is missing; and Frenkel defects, in which an ion leaves its site and enters a lattice hole. Line defects result from missing rows of atoms. Particularly common are edge defects, in which an extra partial plane of atoms is inserted near an edge of the crystal.

PROPERTIES OF LIQUIDS

We take up the liquid phase last because it is the most complex and least understood of the phases of matter. Liquids have properties intermediate between those of the solid and gas phases:

- A sample of liquid has a fixed volume (like a solid) and a molar volume only slightly greater than that of the solid. However,
- A liquid conforms to the shape of its container, like a gas, and
- A liquid flows readily, like a gas.
- Liquids are relatively incompressible. The relative compressibilities of the three phases are in the order g >> l > s. The incompressibility of liquids finds practical use in hydraulic machinery, such as automobile brakes.
- Diffusion is slower in liquids than in gases, but faster than in solids. The relative rates of diffusion in the three phases are in the order g >> l > s.

These properties are readily observed visually. Several other properties of liquids are more subtle:

- Liquids give x-ray diffraction patterns that are diffuse, rather than sharp like those of solids. The pattern implies some order in the molecular arrangement, but less than in the solid.
- Liquids exhibit surface tension.
- Liquids exhibit viscosity.

- Liquids in closed containers evaporate to an extent. The resulting vapour exerts a pressure called the vapour pressure, P_{vap}, of the liquid. The magnitude of P_{vap} for a particular liquid depends only on temperature, not on the container size or the amount of liquid present. Liquid and vapour exist in equilibrium.

We begin with a molecular model for the liquid phase, based on its observable properties.

LIQUID PHASE

The small molar volume and incompressibility of liquids suggest that the molecules are close together. However, flow and the lack of a fixed shape indicate molecular mobility, despite the closeness. A feature of the model that is not evident from the figure is that the molecules are in constant motion, as in the gas and solid, colliding frequently with neighbors. As for the gas and the solid phases, the kinetic energies of the particles of the liquid are described by the Maxwell-Boltzmann distribution, and the average KE per particle is 3/2 kT. Each molecule is fairly closely surrounded by other molecules, but there are definite gaps (holes) in the structure. Molecules use these gaps to slip and slide easily past one another, which manifests macroscopically in the ability to flow.

There are regions in the liquid that are quite ordered, similar to the solid, but the regions are constantly shifting position as molecules move to close gaps and open new ones. The gaps prevent the order from extending over long distances. Thus liquids have short-range order, in contrast to the long-range order of crystalline solids. The gaps allow diffusion of a molecule through the liquid, but frequent collisions with neighboring molecules makes diffusion slow.

Contrast this with the rapid diffusion in the gas phase, where a molecule travels a long distance between collisions; and the extremely slow diffusion of the solid phase, where a molecule is locked into its lattice location. Diffusion rates in solid, liquid, and gas are best understood in terms of the mean free path, the average distance travelled by a molecule between collisions. The mean free path in the solid is virtually zero. In the liquid, it is a fraction of the molecular diameter. But in the gas, the mean free path may be 10-100 molecular diameters, depending on gas pressure.

PHYSICAL PROPERTIES

Indicates regions of short-range order in the liquid. Within these regions, molecules are almost close-packed. However, the gaps allow fluctuation in the distance of separation of the molecules. The nearly close-packed regions of the liquid diffract x-rays according to the Bragg equation, n = 2dsin*theta*. Fluctuations in d, a consequence of the gaps, cause fluctuations in *theta*, giving x-ray patterns that are diffuse, or smeared out.

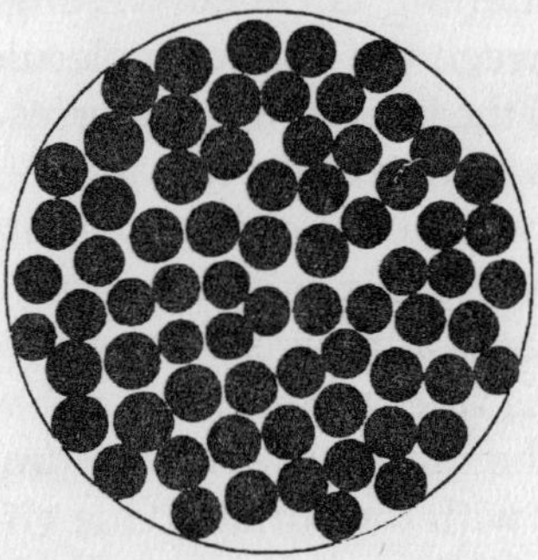

SURFACE TENSION

This is a consequence of intermolecular forces. Consider a molecule at the surface of a liquid. In contrast to a molecule in the bulk, which is attracted equally from all directions, the surface molecule suffers an imbalance of forces because there are no molecules above it. The net force on it is directed toward the interior of the liquid. This pulling in of the liquid surface is called surface tension, symbolized . The surface molecule has higher potential energy than a bulk molecule because of the lesser force on it.

To minimize energy, the liquid adopts a shape that minimizes the number of surface molecules. The shape with the smallest surface area-to-volume ratio is the sphere. Thus surface tension explains the spherical shape of raindrops and water droplets on a freshly waxed car: the spherical shape minimizes potential energy. It is a safe prediction that manufacture of ball bearings will someday be carried out in space, where gravitational forces will not distort the sphericity that surface tension imposes on a drop of molten metal.

As temperature increases, molecular kinetic energy can overcome the attraction of intermolecular forces, and surface tension decreases according to equation.

$$z = a - bT$$

The constants a and b are characteristic of a particular liquid. For ethanol, $a = 24.05 \times 10^{-3}$ J/m^2 and $b = 0.0832 \times 10^{-3}$ J/m^2-°C. A drop of water on a clean glass surface spreads and flattens, forming a film on the surface. This phenomenon of spreading and flattening is called wetting. Wetting occurs because forces between the surface molecules and molecules of the liquid exceed the intermolecular forces within the liquid. The liquid spreads to maximize the area of contact with the surface. Forces between unlike molecules are called adhesive forces; those between like molecules are called cohesive. Wetting occurs because adhesive forces exceed cohesive forces.

If the reverse is true, a liquid will not wet a surface, but instead will sit on the surface in spherical droplets to minimize the area of contact with the surface. One visible consequence of wetting is the formation of a meniscus at the top of a column of liquid in a tube, such as a buret or pipet. The menisci formed by water and mercury. The adhesive forces between the glass surface and water molecules are stronger than the cohesive forces between water molecules. Consequently, water achieves a maximum surface area of contact with the glass by bending up at the edges and sagging in the centre. The reverse is true for mercury, which minimizes the surface area of contact with the glass and forms a spherical surface shape to maximize cohesive forces.

The phenomenon of capillary action results from competition between surface tension and wetting. When one end of a capillary tube (a glass tube with inner diameter of approximately 1mm) is submerged in a liquid, a difference in height develops between the levels of liquid in the capillary and in the bulk liquid. A liquid that wets glass rises in the tube; a liquid that does not wet glass falls, called capillary action, for water and mercury. As water wets the inside wall of the capillary, the surface area of the water is increased.

However, surface tension tends to minimize the number of molecules at the surface, and causes the water to move up the tube, against the force of gravity, to offset the surface area increase resulting from wetting. The result is a capillary rise. For mercury, adhesive forces are weak, so the level in the capillary falls to minimize contact with the glass surface, and the surface of mercury in the capillary is pulled into a sphere by the strong cohesive forces. The result is capillary fall. Capillary action is important in the operation of blotters and sponges and is the mechanism by which water penetrates soils.

VISCOSITY.

The viscosity of a liquid is a measure of its resistance to flow. A mobile liquid is one that flows readily (e.g., water). A liquid that flows with difficulty (e.g., molasses) is said to be viscous. Liquids composed of small, roughly spherical molecules are mobile, because the molecules can roll over one another smoothly, like ball bearings. Viscous liquids are composed of long, string-like molecules that easily become tangled during flow.

For example, the molecules of high viscosity motor oils consist of long chains of carbon atoms that intertwine and impede flow of the liquid. An extreme example of molecular tangling occurs in silly putty, a polymer that appears to be a solid but is in reality a liquid in which molecular tangling is so severe that hours are required for the putty to spontaneously change shape to that of the container. Like surface tension, viscosity decreases (flow becomes easier) as temperature increases due to more rapid thermal motion of the molecules.

VAPOUR PRESSURE.

Vapour pressure is the most fascinating physical property of liquids. It provides remarkable insight into the laws governing the behaviour of physical systems. We begin with a purely experimental approach to the phenomenon. The box is attached to a closed-end manometer and is connected via a stopcock to a vacuum pump.

A thermometer is inserted in the box and there is a septum-capped port (opening) that can be penetrated with a syringe needle.

The stopcock to the vacuum pump is opened and the air is pumped out until the two arms of the manometer are at the same level. The stopcock is then closed.

This is the initial situation, state 1, shown at the top left of the figure. Several experiments will now be performed in sequence.

Vapour pressure can be summarized as follows:

- A liquid in a closed container evaporates to some extent. The resulting vapour exerts a characteristic vapour pressure, P_{vap}.
- The value of P_{vap} is independent of the container size and the amount of liquid, provided that the amount exceeds a critical minimum.
- P_{vap} increases steeply with T.
- P_{vap} is a function only of the chemical identity of the liquid and the temperature.
- Liquids have a natural tendency to evaporate to some extent to form vapour.

The thought experiments, which can be verified readily in the laboratory, provide much descriptive information about the vapour pressure phenomenon. But the questions of why liquids tend to evaporate, and why they establish characteristic vapour pressures.

DISSOLVING POWER OF LIQUIDS.

Liquids are invaluable to chemists as solvents. They are able to accommodate molecules of solute readily within the gaps in their structures to produce solutions that are themselves liquids.

Such solutions are readily poured or otherwise transferred from one container to another; are conveniently measured out quantitatively using volumetric apparatus; and provide a uniform medium in which a chemical reaction can occur. The unique ability of liquids to act as solvents is due to their dissolving power — the molecular motion of the liquid dislodges ions or molecules from the surface of crystals of a solid, then accommodates the separated ions or molecules in gaps. In this section we briefly discuss the dynamics of the dissolution process at the molecular level.

A microscopic view of two adjoining faces of a crystal of NaCl. A Na^+ ion on a face of the crystal differs from those in the interior because it experiences forces of attraction from only 5 nearest-neighbour Cl^- ions, rather than 6.

Facial Cl^- ions are in a similar situation. Such ions are therefore in shallower potential wells than ions in the interior of the crystal and should be more easily dislodged. Edge ions, such as the Na^+ ion labelled "a" in the figure, are attracted by only 4 nearest neighbors, and should be even more readily dislodged than face ions. Water molecules from the solvent continually bump against the faces and edges of the solid crystal.

As they approach the crystal, their dipoles become oriented due to interactions with the face ion that they approach. Water molecules approaching cations have the O atom oriented toward the crystal face; those approaching anions put a H atom forward.

When a rapidly moving water molecule strikes a face or edge ion, sufficient kinetic energy can be transferred to dislodge the ion, which moves away from the crystal into a gap in the solvent structure (e.g., Na^+ ion "b" in the figure). It becomes immediately surrounded by a cage of water molecules, with oxygen atoms nearest the Na^+ ion to maximize ion-dipole forces.

The solvent cage insulates it from the crystal and makes return to the crystal face unlikely. Since edge ions are more readily dislodged than facial ions, and because they can be approached by solvent from several directions, dissolution occurs more rapidly at crystal edges than faces. The initially sharp edges of a dissolving crystal become rounded as dissolution proceeds.

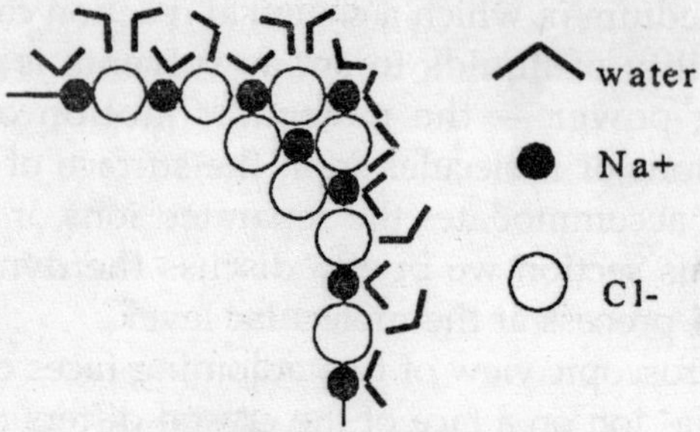

PHASE TRANSFORMATIONS.

A phase transformation is a conversion of a pure substance from one phase to another. There are 6 types, each involving two of the three phases:

- *Melting (fusion)*: Conversion of solid to liquid.
- *Freezing*: Conversion of liquid to solid; the reverse of melting.
- *Vaporization*: Conversion of liquid to vapour (gas). Its reverse is
- *Condensation*: Conversion of vapour to liquid.
- *Sublimation*: Conversion of solid to vapour. Its reverse is
- *Deposition*: Conversion of vapour to solid.

The phase changes are illustrated in Figure. Since the solid and liquid phases have small molar volumes due to close packing of molecules, they are referred to as condensed phases.

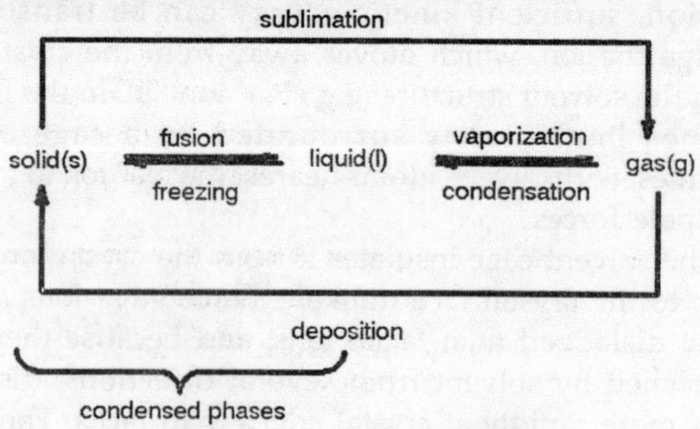

Fig. Phase Transformations

The two variables that determine the phase in which a

substance exists are the temperature and the pressure to which the substance is subjected. Phase transformations may therefore be accomplished in two limiting ways: by changing the temperature of the substance while keeping it under constant pressure; or by changing the pressure on the substance while keeping it at constant temperature. The first approach is the more familiar of the two.

We make three assertions about temperature-induced phase changes:

- At a constant pressure, a phase transformation occurs at a constant temperature that is characteristic of the substance.
- To convert a given amount of substance from solid to liquid to gas requires addition of energy. This is usually accomplished by adding heat. The energies of the phases are therefore in the following order: $E_s < E_l < E_g$
- Conversion from solid to liquid to gas decreases the order (degree of organization) of the substance.

To clarify these assertions, we take a thought experimental approach to the phase changes of a substance, much as we did for vapour pressure. The results of these thought experiments can be readily verified in the laboratory. We start with a quantity of solid water (ice) in a syringe, state 1. The initial temperature of the ice and the inside of the syringe is -10°C. We plan to slowly add energy to the ice by controlled heating, which can be done by submerging the syringe in a large antifreeze bath with a hot plate under it. As we slowly add heat, we monitor the temperature of the water in the syringe with a thermometer inserted through an air-tight seal in the syringe. We begin at state 1.

Process: Turn on the hot plate and the bath stirrer. As the ice warms slowly from -10 to 0°C, nothing seems to happen. Temperature changes slowly and steadily. When the ice reaches 0°C, it begins to melt, and we see some liquid water form in the syringe.

Process: We continue to gently heat the bath. The ice gradually melts, but the temperature of the ice-water mixture remains constant at 0°C, despite the addition of heat. The temperature of the bath rises above 0°C, but as long as both ice and liquid water are in the syringe, T in the syringe remains at 0°. State 3 is reached when the last ice crystal melts. All water in the syringe is liquid, at T = 0°C.

Process: Once the ice has melted, the temperature of the water in the syringe begins to rise. As we slowly warm the bath, the water temperature slowly rises. Nothing happens in the syringe other than a slight increase in the volume of the liquid until T = 100°C. Then water begins to vaporize. This is the situation at state 4.

Process: We continue to slowly add heat to the bath and observe that temperature in the syringe remains constant at 100°C as long as both liquid and vapour are present, even though the bath temperature rises above 100°C. We also observe that as vapour forms, it pushes the plunger out. After all liquid has vaporized, at state 5, the temperature of the water in the syringe once more starts to rise.

Process: Continue heating the bath to T = 110°C. The temperature of the vapour in the syringe rises from 100° at state 5 to 110° at state 6, while its volume increases according to the ideal gas law. The pressure of the water vapour is maintained constant at 1 atm by movement of the syringe plunger.

These experiments verify the assertions at the beginning of the section. First, phase changes occur at constant T. This is shown by processes 2 and 4. The fixed, characteristic temperature at which a substance melts is called the melting point (s to l) or freezing point (l to s), and is symbolized T_f. The normal melting point is the temperature of melting under a pressure of 1.00 atm. For water, this is 0°C. The characteristic temperature at which a substance boils under 1.0 atm pressure is called the normal boiling point. For water, this is 100°C. Second, process 2 shows that an input of energy is required to convert a solid to liquid. The same is true for conversion of liquid to gas (process 4). For each of these processes, the change in energy, DE (E_{final} - $E_{initial}$) is positive. Processes 2 and 4 may be represented as follows:

$$H_2O(s),\ 0°C \rightarrow H_2O(l),\ 0°C\ [DE > 0]$$
$$H_2O(l),\ 100°C \rightarrow H_2O(g),\ 100°C\ [DE > 0]$$

The process in equation occurs at constant T. Therefore the change in kinetic energy of the molecules is 0 (recall that an average molecule has KE of 3kT/2 at temperature T, regardless of phase). DE is therefore a change in potential energy, DPE. Thus the PE of liquid is higher than that of solid water at 0°C. Similarly, the PE of vapour is higher than that of liquid water at 100°C. Combining these conclusions gives equation:

$$PE_g > PE_l > PE_s \text{ at any T}$$

DPE > 0 for sublimation, fusion, and vaporization; DPE < 0 for the reverse processes. Third, the amount of disorder at the molecular level increases in our sequence of processes (for reasons that will become clear later, we phrase this in terms of disorder rather than order). With addition of thermal energy (heat), the molecules convert from the long-range organized arrangement of the solid to the completely chaotic and random molecular distribution of the gas. Even warming the gas from 100 to 110°C decreases the order, because the volume available for the molecules to bounce around in increases. This conclusion is summarized in equation.

disorder s < disorder l < disorder g

PHASES AND POTENTIAL WELL

The origin of the potential well in terms of electromagnetic forces within or between the fundamental units of substances. We concluded that forces are strongest and potential energy is minimized when fundamental particles form an orderly close-packed arrangement, and we recognized this as the situation in the solid phase. Heating the solid phase increases the kinetic energies of the particles of the solid phase until they are able to partially overcome the close-packed forces and separate somewhat.

At this point, solid melts to liquid. Separation has increased the potential energy of the particles. Further heating eventually gives the particles sufficient kinetic energy that they may escape the forces in the liquid into the space above the liquid; i.e., vaporization occurs. In the vapour phase, molecules are so widely spaced that forces between them are negligible and potential energy has increased to zero. We thus see that phase conversion from solid to gas can be viewed in terms of particles climbing out of the potential well. Similarly, in the reverse conversion, particles fall into the potential well. The decrease in potential energy of the molecules is released as heat.

Chapter 3

Thermodynamics

FIRST LAW

From experience, phase changes of pure substances involve energy changes. For example, input of heat is necessary to melt ice, or to boil water. Energy is released as heat when steam condenses or water freezes. In this chapter we will lay the foundation for quantitative discussion of energy changes in chemical reactions and in physical processes, such as changes of phase.

The formal structure for discussion of energy is a portion of the discipline of Thermodynamics. Thermodynamics began as the study of the exchange of heat and work between an engine and its surroundings. From these humble beginnings has grown a branch of science so powerful that it can tell us whether or not a process of interest can occur, and to what extent. We now begin a study of thermodynamics. We will return to a discussion of phase changes and build upon the energy concepts developed here.

Scientists are concerned with the interaction between a system and its surroundings, which occurs across a boundary between them. A system may be changed from one state to another by a process, which usually involves an exchange of energy between system and surroundings. The emphasized words are important. They occur frequently in any discussion of science, and should mean the same thing to all who use them. We define each of these words and a few others below:

- *System*: The part of the universe in which we are interested.

- *Surroundings*: Everything else.
- *Boundary*: The dividing line between system and surroundings, usually real, but sometimes imaginary.
- *State (of a system)*: A particular condition in which a system is found; "condition" usually refers to specific values of the variables, P, V, T, and quantity in moles of each component of the system.
- *Closed system*: A system with a boundary permeable to energy but not to mass.
- *Internal energy (E)*: All of the energy possessed by the system.
- *Process*: A path by which the system proceeds from one state to another.
- *Isothermal process*: A process at constant temperature.
- *Adiabatic process*: A process in which no heat can flow across the system boundary.

Systems frequently change energy in processes, and it is these changes that we actually measure. The change in energy of a system in a process will be written ΔE, where the symbol, Δ, means "change." The change in energy is the difference between the energy after the process (final value) and the energy before the process (initial value):

$$\Delta E = E_f - E_i$$

Any process involves two energy changes, one for the system, the other for the surroundings.

These are always equal and opposite in sign:

$$\Delta E_{sys} = -\Delta E_{surr}$$

Energy lost by the system must be gained by the surroundings, and vice versa, because energy is a conserved quantity: the total energy of the universe is constant. Generally, we will be more interested in ΔE_{sys} than in ΔE_{surr}; unless otherwise stated, the symbol ΔE will refer to the energy change of the system.

As we have seen, energy is of only two types: kinetic (motion) and potential (position). The energy, E, of a system is the sum of its kinetic and potential energies:

$$E = KE + PE;$$

the change in E is the sum of changes in KE and PE:

$$\Delta E = \Delta KE + \Delta PE.$$

Kinetic energy is possessed by a moving baseball (translation), a spinning top (rotation) and a spring that is repeatedly stretching and contracting (vibration). Molecules undergo analogous motions.

Potential energy is possessed by a rock on a mountain, a compressed spring, a positive and a negative charge separated by some finite distance (as in a chemical bond) and by mass (through the Einstein equation, $E = mc^2$). Potential energy results from forces. Important for chemists is the PE associated with chemical bonds. When bonds are broken, the Coulomb force of attraction of two nuclei for the bonding electrons must be overcome, and PE increases. When bonds are formed, force of attraction is increased and PE decreases.

Thus,

$$\text{H-H} \rightarrow 2\ \text{H}\ [\text{DPE} > 0]$$
$$2\ \text{Cl} \rightarrow \text{Cl}_2\ [\text{DPE} < 0]$$

Energy changes in chemical reactions are primarily potential energy changes that result from breaking bonds and making new ones.

HEAT AND WORK

Suppose we desire to change the internal energy of a system. How can we do it? Surprisingly, there are only two ways: we can transfer heat to or from the system; or we can arrange for work to be done on or by the system. All of thermodynamics is built upon the measurement of heat and work. The quantity of heat involved in a process is represented by the letter, q. We will adopt the convention that q is positive when heat is added to a system, and negative when it is removed from a system. The amount of work involved in a process is represented by the letter, w. We will adopt the convention that w is positive when work is done on a system by its surroundings, and negative when work is done by the system.

If an amount of heat q is added to a system, and an amount of work w is done on it, the (internal) energy of the system changes by the amount DE. This very simple and important equation is called the First Law of Thermodynamics. It is a statement of the law of energy conservation.

$$\Delta E = q + w$$

Before proceeding further, we discuss a point that is probably obvious but that is nonetheless important. Although our focus will normally be on the heat added to the system (whether positive or negative) and the work done on the system by the surroundings (whether positive or negative), we will on occasion prefer to view things from the perspective of the surroundings.

That is, we will be interested in the heat added to the surroundings, or the work done on the surroundings by the system. In all processes, the heat added to the system and to the surroundings are equal and opposite (that is, if heat enters the system, it must leave the surroundings, and vice versa). Similarly, in all processes, the work done on the system and by the system are equal and opposite (that is, if the system does work on the surroundings, the surroundings does an equal and opposite work on the system).

Thus,

$$q_{sys} = -q_{surr}$$
$$w_{on\ sys} = -w_{by\ sys}$$

Heat and work are familiar yet elusive concepts. We may define heat as energy that "flows" due to a temperature difference. The spontaneous movement of heat is always from high to low temperature. For example, heat flows from a hot cup of coffee to the cooler surroundings, but never in the reverse direction. The quantity of heat added to a system can be readily determined from the temperature change of the system and its heat capacity, C:

$$q = C^*\Delta T$$

The heat capacity of a system is defined as the quantity of heat that must be added to the system to increase its temperature by 1°C (or 1 K).

For a pure substance, the value of C depends on the amount of substance being heated (or cooled). Values of the heat capacity per gram (the specific heat, C_s) or heat capacity per mole (the molar heat capacity, C_m) are therefore tabulated.

Table gives the specific heats and molar heat capacities for a number of common substances.

Substance	C_s, J/g-K	C_m, J/mole-K
$H_2O(l)$	4.184	75.376
Fe(s)	0.449	25.1
$O_2(g)$	0.9172	29.35
NaCl(s)	0.8641	50.50
Al(s)	0.901	24.3

From the molecular viewpoint, heat is related to the changes in the random thermal motions of molecules. The molecules of a system that is hotter than its surroundings have higher average kinetic energy than the molecules of the surroundings. These "hot" molecules collide with the boundary of the system and transfer some of their kinetic energy to the molecules of the boundary. Molecules in the surroundings in turn receive kinetic energy by colliding with the boundary molecules. In this way, the excess thermal (kinetic) energy of the system molecules is transferred across the boundary to the molecules of the surroundings.

The average KE of the system molecules decreases and that of the molecules in the surroundings increases in this way, until system and surroundings are at the same temperature. Once this situation has been reached, there will be no further tendency for kinetic energy to be transferred one way or the other; a state of thermal equilibrium has been reached. Thus heat flow is the result of a difference of molecular kinetic energies on two sides of a boundary.

We define work as the energy expended when a force moves matter through a distance. Mathematically, it is the product of the force, F, exerted and the distance, r, moved.

$$w = F * r$$

This equation is correct only when the force remains at the same value over the entire distance. Since this is frequently and generally not the case, we replace equation with the differential equation,

$$dw = F * dr$$

Here dw is the infinitesimal amount of work done when

force F moves a mass through the infinitesimal distance, dr. The distance is short enough that F may be considered constant. To obtain the total work done over the full distance, r, which is the sum of the infinitesimal distances, dr, we add up the infinitesimal amounts of work. This is done by integration of equation.

$$w = \text{Integral}[F\ dr]$$

Equation is the general expression for work in any situation. We now wish to tailor it to a form that is more useful for chemical applications. Since the work done when chemical reactions take place is the result of expansion or contraction of the system volume against a pressure exerted by the surroundings, we must translate equation into pressure-volume terms.

The figure shows a gas confined in a cylinder by a piston of area A. The surroundings exerts a pressure, P_{ext}, on the piston. We will calculate the work done on the system when the external pressure contracts it an amount ΔV_{sys}, from V_i to V_f ($V_f < V_i$), assuming that the external pressure remains constant throughout the contraction. During the contraction, the piston moves in a linear distance, d. Since the piston moves in the same direction that the external pressure is pushing, the work done by the surroundings (on the system) is positive. The volume change of the surroundings during the process is ΔV_{surr} = Ad. But the volume change of the system is the negative of this, so that Ad = $-\Delta V_{sys}$. The force exerted by the surroundings is F = P_{ext}A, since pressure is force per unit area. It follows that the work done in the process is,

$$w = Fd = (P_{ext}A)(-\Delta V_{sys}/A) = -P_{ext}\Delta V_{sys}$$

We have successfully expressed the work in terms of the outside pressure and the volume change of the system. There are two things to note about this equation. First, the system has been contracted by the surroundings, so work has been done on it. According to our convention, w should be positive. However, ΔV_{sys} is < 0. The - sign on the right side insures that a positive work is calculated when the system is compressed. Second, P_{ext} need not be constant throughout the contraction process. We must therefore sum over infinitesimal amounts of work as we did to obtain equation. The final result for the mechanical work of expansion, important for chemical processes, is equation.

$$w = -\text{Integral}[P_{ext} dV_{sys}]$$

Two things are noteworthy about this equation. First, it says that there can be no work done unless the system changes volume; that is, a boundary of the system must move. Second, if P_{ext} is constant throughout a process, then integration is simple and $w = -P_{ext}\Delta V_{sys}$.

At the molecular level, work results because random molecular collisions with a movable system boundary transfer momentum to the boundary and push it back against an external force. The motions of the system molecules are completely random, as always. But their random motion is converted to an organized and macroscopically observable motion of the boundary that can be used to do something useful in the surroundings. Work is quite different from heat, which "flows" because random molecular motions on one side of the boundary are converted to random molecular motions on the other side. Heat never leaves the realm of randomness; work does.

Work is done on a system by the surroundings when a moveable system boundary is pushed in by the surroundings. The unidirectional boundary movement transfers energy to the system molecules that collide with it, much as a ping-pong ball is accelerated by a moving paddle. This tends to increase the average kinetic energy of the system molecules, and the system temperature goes up; the system warms when work is done on it if no heat is transferred.

STATE AND PATH FUNCTIONS

Examples illustrate two isothermal gas expansions, with the initial and final states of the system the same in both cases. It is instructive to compare the results:

Isothermal expansion against constant P_{ext}:

$$\Delta E = 0$$

$$w = -2.0 \text{ L-atm}$$

$$q = 2.0 \text{ L-atm}$$

Isothermal free expansion:

$$\Delta E = 0$$
$$w = 0$$
$$q = 0$$

The amounts of heat and work that "flow" during a process connecting specified initial and final states depend on how the process is carried out. In contrast, the change in the energy of the system is the same in both expansions, independent of how the process is performed. Quantities that depend on the path followed between states are called path functions.

Quantities whose values are independent of path are called state functions. The total change in the value of a state function for any cyclic process (a process in which the initital and final states of a system are the same) is zero. We say that state functions are conserved during any complete cycle. Path functions are not conserved. In thermodynamics, state functions are represented by upper case (capital) letters; e.g., P, V, T, and E. Path functions are represented by lower case letters; e.g., q and w.

In light of this discussion, the first law may seem odd, because it expresses the change in a state function as the sum of two path functions. Although heat and work are both path-dependent quantities, their sum is not. That this should be true is not obvious. That it is true has been verified in all known applications of the first law of thermodynamics.

ENTHALPY

Enthalpy, symbolized H, is the most important energy quantity for discussion of chemical reactions. It is defined by equation:

$$H = E + PV$$

Enthalpy is a state function of the system because E, P, and V are state functions. It is obviously an energy quantity because both terms on the right side of equation have energy units.

At this stage of our development, though, it is not clear why we need a new state function, or why it should have the form of equation. We attempt to justify the form before moving on. From the first law of thermodynamics it is obvious that if a process is carried out at constant volume (so that $\Delta V_{sys} = 0$), no work is done and $\Delta E = q$. This is an important result; it tells us that if we measure the heat added to a system in a constant volume process, we immediately know the change in energy of the system. To emphasize the constant volume restriction on the equation, it is usually written:

$$\Delta E = q_V$$

Unfortunately, although the equation is useful, it is not convenient to carry out processes at constant volume. This requires using completely closed containers, sealed off from the outside, which makes isolation and study of the product of a chemical reaction difficult. In practice, chemists do most of their work under conditions of constant pressure, and let the volume of the system change as it will.

In most cases, chemical reactions are carried out in containers that are open to the atmosphere, and the constant pressure is atmospheric pressure.

An energy state function whose change is measured by the heat added to the system in a constant pressure process would therefore be more useful than E for the analysis of chemical reactions. Such a state function, which we might call H to suggest its connection with heat, would then satisfy equation:

$$\Delta H = H_f - H_i = q_P$$

What form should the new function H have? We can answer this by applying the first law to the constant pressure process:

$$\Delta E = q + w = q - \text{Integral}[P_{ext} dV_{sys}]$$

for a constant pressure process, $P_{ext} = P_{sys}$ = constant. We may therefore replace P_{ext} with P_{sys} and take the pressure out from under the integral sign to obtain equation:

$$\Delta E = q - P\Delta V_{sys}$$

Expanding the terms and solving for q gives: $q = E_f + PV_f - (E_i + PV_i)$

Now notice that in equation the difference between the final and initial values of the quantity, E + PV appears. We can make equation look like equation by substituting H = E + PV.

The final result is equation: $DH = q_P$

The subscript P is added to remind us of the restrictions on the truth of the equation. This explains why enthalpy is defined as it is in equation, and suggests the following statement: the enthalpy change of a system for a process carried out at constant pressure is the amount of heat absorbed by the system during the process. Based on equation, we may think of enthalpy as the energy of the system plus the energy of interaction of the system with the surroundings, represented by the PV term. When P is constant, this interaction energy is the work done by the system on the surroundings.

MOLECULAR ENERGY

The internal energy of a system is the total of the energy possessed by its atoms and molecules. We have discussed the types of molecular energy in terms of equation. We will now apply that equation to three systems of increasing molecular complexity, to show ultimately that the most important contribution to molecular energy is the potential energy of chemical bonds.

First, consider a system consisting of an ideal gas. There are no intermolecular forces between molecules of an ideal gas, so there is no intermolecular potential energy; and there is no internal molecular structure to worry about. The only energy that such a system has is the translational kinetic energy of the molecules as they move about in the container. For 1 mole of an ideal gas at 298 K, the total energy is:

$$E = KE_{trans} = 3RT/2 = 3.7 \text{ kJ}$$

We now increase the complexity of our system slightly by considering a real monatomic gas. Monatomic means that the "molecules" are single atoms. Examples of real monatomic gases are the noble gases, He, Ne, Kr, Xe, and Rn. Atoms of these gases exert weak (but non-zero) intermolecular forces of the instantaneous dipole-induced dipole type on one another, so there will be a small potential energy in such a system in addition to the translational kinetic energy at 298

K. In addition, real atoms have internal structure (i.e., nuclei and electrons). We will ignore the energy associated with this structure.

For real monatomic gases, then:

$$E = KE + PE_{inter} \text{ (small because forces weak)}$$
$$= 3.7 + (-1.2) \text{ kJ} = 2.5 \text{ kJ}$$

The value of potential energy used here is for the noble gas, Ar. It is negative because we follow the convention established potential energy is 0 when the atoms or molecules are far enough apart that the forces fall essentially to 0, and decreases from zero at smaller distances as the attractive forces take effect.

We now make one more increase in complexity and consider a real polyatomic gas. In this case, each gas molecule contains 2 or more atoms. Examples are HCl (hydrogen chloride), NH_3 (ammonia), and CH_4 (methane). The total energy of such a molecule, again ignoring the internal atomic structure, is in equation.

$$E = KE_{trans} + KE_{rot} + KE_{vib} + PE_{vib} + PE_{inter} + PE_{bonds}$$

This equation bears some resemblance to equation. E_{spin} and E_{nuclei} have been omitted because they are not involved in chemical reactions, and therefore will not change. E_{trans} and E_{rot} have been explicitly acknowledged as kinetic energies. E_{vib} has been divided into kinetic and potential portions (the energy of vibration interconverts between potential and kinetic as the vibration moves from either extreme (stretched or compressed), where there is no motion, to the point where there is no tension in the spring). PE_{inter} acknowledges that molecules exert attractive forces on one another, just as do the atoms of argon. Finally, PE_{bonds} is the potential energy resulting from the chemical bonds within the molecules, holding the atoms together.

Equation is evaluated for HCl in equation:

$$E = 3.7 + 2.5 + 1.2 + 1.2 - 1.2 - 431 \text{ kJ}$$
$$= -422 \text{ kJ}$$

The first 5 terms of equation are similar in magnitude and small. The dominant term is clearly the potential energy of chemical bond formation. We will generalize from this one specific example: the internal energy of a system consisting of real polyatomic gases is primarily the potential energy of chemical bonds.

When we consider E for chemical reactions in the next section, we will equate it to the change in bond potential energy, and assume that the remaining (small) terms in equation are very nearly the same for reactants and products.

APPLICATIONS OF CHEMICAL REACTIONS

A chemical reaction in its most general form is presented in equation.

$$\text{Reactants} \rightarrow \text{Products } [q, \text{DE (PE)}, \text{DH}]$$

In thermodynamic terms, the reaction is a process in which a system proceeds from an initial state (the reactants) to a final state (the products). During the process, an amount of heat, q, is added to the system. The system undergoes a change in internal energy, DE, and a change in enthalpy, ΔH. As we have seen, both ΔE and ΔH can be related to q under certain conditions. A reaction for which $q > 0$ (i.e., in which heat is transferred from surroundings to system) is endothermic ("endo" means "in"; thus, heat in); one for which $q < 0$, exothermic ("exo" means "out"; hence, heat out).

If the reaction is done at constant pressure, $q = \Delta H$. If the reaction is done at constant volume, $q = \Delta E = \Delta PE_{bonds}$. We state without proof that ΔH and ΔE are approximately equal for chemical reactions, and both provide a good measure of the change in potential energy that occurs when bonds in the reactants break and bonds in the products form. In this text, we will not worry about the small difference between ΔH and DE. In more advanced courses, the distinction between the two quantities will be made clear.

HESS'S LAW

It should come as no surprise that enthalpy, a form of energy, is conserved in chemical reactions. This is something that is now well-established and that we rely upon in chemistry. However, in the early application of thermodynamics to chemical reactions, this was not known. It was put on a firm experimental basis by Hess. We will introduce the ideas generally, then apply them to a particular example.

Suppose that we are interested in the enthalpy change, DH_{unk} for the chemical reaction

$$A + 2B \rightarrow C\ [\Delta H_{unk}]$$

but for some reason are unable to determine it experimentally (by carrying out the reaction and measuring the heat produced or absorbed). Suppose, however, that we can measure the enthalpy changes for two related reactions:

$$(1)\ A + B \rightarrow D\ [\Delta H_1]$$
$$(2)\ D + B \rightarrow C\ [\Delta H_2]$$

The reactions are related, because when they are added together and substances common to both sides cancelled, equation results. It would be useful if the three enthalpy changes were also related in this way; i.e., $\Delta H_{unk} = \Delta H_1 + \Delta H_2$. We can use an energy level diagram to show that this should be true. Figure shows three states plotted on such a diagram.

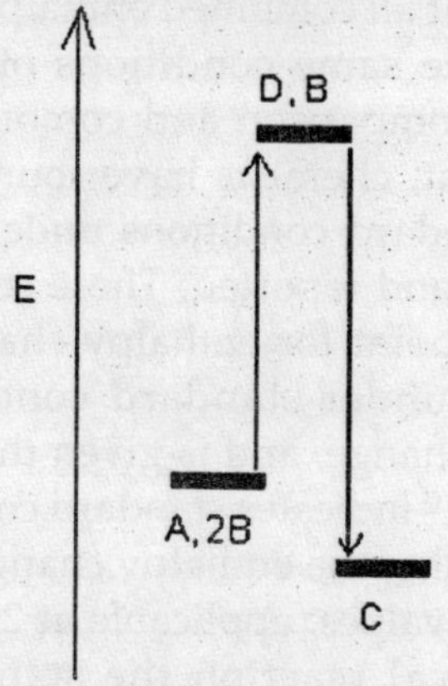

Fig. Energy Level Diagram

The initial state for the unknown reaction is A, 2B; the final state for the unknown reaction is state C. The intermediate state on the diagram is the initial state for reaction (2). ΔH_1, ΔH_2, and ΔH_3 are shown as arrows on the diagram (strictly speaking, the reaction in the diagram is

$$A + 2B \rightarrow D + B$$

However, cancellation of B gives reaction (1). It is clear

that the arrow representing ΔH_{unk} is the vector sum of the arrows representing ΔH_1 and ΔH_2. This demonstrates what we wanted to show.

Hess measured the enthalpies for many series of related reactions and found that in all cases, enthalpy was conserved. He summarized his work in a simple statement, which is paraphrased below:

Hess's Law: if several reactions can be added or subtracted to give a desired reaction, their enthalpy changes can be combined in the same way to give the enthalpy change of the desired reaction.

Stated another way the law says that no matter what pathway we take from one state to another, the net change in enthalpy will be the same. Enthalpy is a state function.

ENTHALPY OF REACTION

A restriction on the use of Hess's Law that we did not mention above is that all combined enthalpies must have been measured under the same conditions of temperature and pressure. Because comparison and combination of enthalpy changes are frequent, chemists have found it convenient to specify a set of standard conditions under which enthalpies are to be measured and reported. These conditions provide a common reference point for enthalpy changes. An enthalpy change measured under standard conditions is called a standard enthalpy change, and is given the symbol DH^o.

The superscript ^{o} indicates standard conditions. Although it is not indicated, all of the enthalpy changes in the examples above are standard values, applicable at 298 K.

For the chemical reaction the standard enthalpy of reaction, ΔH_R^o (the subscript R stands for "reaction"), is defined as the enthalpy change that occurs when the reaction is carried out under standard conditions – i.e., with each reactant and product in its standard state.

$$aA + bB \rightarrow dD + fF$$

The standard state of a pure substance is the form (solid, liquid, or gas) in which the substance is most stable when subjected to 1 atm pressure at a specified temperature.

Solutions are in the standard state when all solutes are present at 1.0 M concentration, again at a specified temperature. In most cases, the specified temperature is 298 K (for convenience), but it need not be. For example, hydrogen is most stable as a diatomic gas at 1 atm pressure and 298 K; its standard state at 298 K is therefore $H_2(g)$. The standard state of water at 298 K is $H_2O(l)$, but at 398 K is $H_2O(g)$. That of iron(III) oxide is the solid ionic lattice, $Fe_2O_3(s)$, at 298 K. Each substance has only one standard state at a specified temperature, but many non-standard states. Several non-standard states for hydrogen at 298 K are listed below:

- *H(g) (gaseous atoms)*: This state has higher enthalpy than the state in which the same number of atoms is combined as diatomic molecules, because the bonds in the diatomic molecules must be broken to create the unbonded atoms.
- *$H_2(l)$ (diatomic liquid)*: This state has lower enthalpy than the standard state because intermolecular forces lower the PE.
- *$H_2(g)$ at 0.5 bar*: In this case the substance has the correct physical state (gas), but only half of the standard pressure of 1 atm.

The symbol, ΔH_R^o, is used for the enthalpy change of any reaction carried out under standard conditions.

ENTHALPY CHANGES

A vast number of reaction enthalpies has been measured, so many that to tabulate them all would require a compendium of unmanageable size. In addition, such a compendium would be immediately out of date as new reaction enthalpies became known.

Clearly some efficient means of tabulation is desirable, and has been achieved using the so-called standard enthalpy of formation, symbolized ΔH_f^o. The standard enthalpy of formation of a substance is defined as the enthalpy change that occurs when 1 mole of the substance is made from its constituent elements in their standard states.

Example: Write the formation reactions for the substances

$C_2H_5OH(l)$ (ethyl alcohol), $CaCO_3(s)$ (calcium carbonate), $C_2H_2(g)$ (acetylene), and $Br_2(l)$ (elemental bromine).

Solution: We begin by writing 1 mole of the desired substance as the only product of a chemical reaction. For ethyl alcohol, the starting point is,

$$\rightarrow 1\ C_2H_5OH(l)$$

The elements of ethyl alcohol are carbon, hydrogen, and oxygen. Their standard states are C(graphite), $H_2(g)$, and $O_2(g)$. We write these elemental standard states on the left of the equation:

$$C(\text{graphite}) + H_2(g) + O_2(g) \rightarrow C_2H_5OH(l)$$

Finally, we balance the equation by changing only the coefficients on the left (not the right, because we must form 1 mole of the desired product):

$$2C(\text{graph}) + 3H_2(g) + 1/2\ O_2(g) \rightarrow C_2H_5OH(l)$$

The standard enthalpy change for this reaction is by definition $DH_f°(C_2H_5OH)$. In similar fashion we obtain the following formation reactions for the other 3 substances:

$$Ca(s) + C(\text{graph}) + 3/2\ O_2(g) \rightarrow CaCO_3(s)$$

$$2C(\text{graph}) + H_2(g) \rightarrow C_2H_2(g)$$

$$Br_2(l) \longrightarrow Br_2(l)$$

The last reaction is interesting because nothing happens! Elemental bromine is made from itself. The enthalpy change must be zero. The same will be true when any other element in its standard state is made from itself. Thus, the standard enthalpy of formation of an element in its standard state is zero.

As Example.indicates, standard enthalpies of formation may be viewed as chemical potential energies of substances relative to the chemical potential energies of the elements in their standard states.

Rather than keep a huge, running compendium of standard enthalpy changes for arbitrary reactions, chemists tabulate $DH_f°$ values for individual substances at 298 K. From the standard enthalpies of formation, it is possible to calculate the standard enthalpy changes for many reactions involving these substances. The latter enthalpy changes then need not be tabulated. The figure shows a general chemical reaction in which some set of reactants, R, is converted to products, P. We can construct a second pathway for this process involving two steps. In the first step, Reactants are decomposed into their constituent elements in the standard state.

The associated enthalpy change is simply the negative of the sum of the standard enthalpies of formation of the reactants.

The negative sign is required because reactants are being decomposed, not formed. In the second step, these elements are recombined to form the Products. The enthalpy change for this step is the sum of the standard enthalpies of formation of the products. Hess's Law tells us that the net enthalpy changes for the two paths must be the same. It therefore follows that the enthalpy change for the overall reaction is given by equation:

$$DH_R^o = S[DH_f^o(\text{products})] - S[DH_f^o(\text{reactants})]$$

Standard enthalpies of formation for a few common substances are given in Table.

Table. Standard Enthalpies of Formation, kJ/mole

Substance	DH_f^o
CaCO3	–1207.1
CaO	–635.5
CH4(g)	–74.8
C2H2(g)	226.7
C2H4(g)	52.3
C6H12O6(s)	–1273.02
CO2(g)	–393.5
H2O(l)	–285.85
MgO(s)	–601.8

Figure below shows several standard enthalpies of formation plotted on an energy diagram. The energies are plotted relative to those of the elements in their standard states, all found at zero on the diagram. Using a diagram like this, it is possible to use vector addition to obtain the value for DH_R^o of a specified reaction. Such a procedure is the graphical analogy to equation.

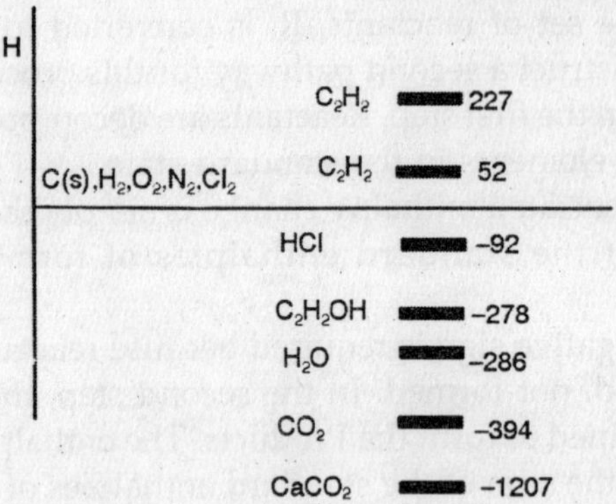

Fig. Standard Enthalpies of Formation

ENTHALPIES OF PHASE TRANSFORMATION

We described a thought experiment in which we began with ice in a syringe at -10°C. We gradually added heat to the system, watching the changes within it, until the final temperature was 110°C and the water was present as vapour. All pure substances undergo temperature-induced conversion from the solid state to the gas state in essentially the same way. The range of temperature involved will vary from substance to substance, but the phase change characteristics do not. Figure shows a plot of the heat added per mole of substance versus the temperature of the substance.

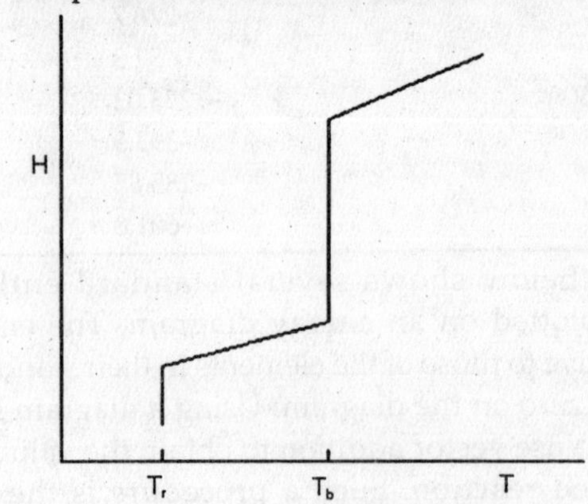

Fig. Heat Per Mole Versus T

The plot is called a heating curve. In region 1 (T_1 to the melting point, T_f) the solid is heated.

The added heat increases the temperature (kinetic energy) of the molecules of the solid. The amount of heat added per mole of substance is linearly related to temperature according to equation:

$$q/n = C_{solid}DT$$

The slope of the plot in region 1 is the heat capacity of the solid, C_{solid}. Since the addition of energy to the system is carried out at constant pressure, $q = \Delta H$.

Thus:

$$\Delta H = nC_{solid}DT$$

When the melting temperature is reached, it is maintained until all solid has melted. Region 2 (T_f) shows that the temperature of the system remains constant despite the addition of heat. In this region, the kinetic energy of the system is constant but the potential energy increases as molecules pull apart against the intermolecular forces; the solid melts.

The energy added in this region is called the heat (or enthalpy) of fusion, ΔH_{fus}. Region 3 (T_f to T_b) is like region 1 except that liquid, not solid, is present, so the slope is C_{liquid}. Region 4 (T_b) shows vaporization at the boiling point, T_b, with addition of the enthalpy of vaporization, ΔH_{vap}.

Finally, region 5 shows heating of the vapour at a rate C_{gas}. Regions 2 and 4 clearly show the system climbing out of the potential well of the solid phase, first melting to liquid, then vaporizing to gas.

For most substances, ΔH_{vap} is larger than ΔH_{fus} because the increase in distance of molecular separation from liquid to vapour is much larger.

The values of ΔH_{fus} and ΔH_{vap} in Figure above apply at the temperatures T_f and T_b, respectively. Just as for chemical reactions, however, it is convenient to tabulate phase change enthalpies for 1 mole of substance under standard conditions at 298 K.

Table shows standard molar enthalpies of fusion, ΔH^o_{fus}, and vaporization, ΔH^o_{vap}, for a number of common substances.

Table. Standard Molar Enthalpies of Fusion and Vaporization, kJ/mole

Substance	ΔH_{fus}^{o}	ΔH_{vap}^{o}
water	6.01	40.7
carbon tetrachloride	2.47	30.00
carbon dioxide	8.33	25.23(s)
methyl alcohol	3.16	35.27
ethyl alcohol	4.60	43.5
ammonia	5.65	23.35

Example: When 1 mole of water boils at 100°C and a constant pressure of 1 atm, 40.67 kJ of heat is absorbed. Calculate ΔH and ΔE for the vaporization process and show that they are similar.

Solution: The equation for the process is,

$$H_2O(l) \rightarrow H_2O(g), \text{ 1 atm}$$

Since the process is done at constant pressure,

$$\Delta H_{vap} = q = 40.67 \text{ kJ.}$$

The definition of enthalpy can be used to obtain ΔE:

$$H = E + PV \text{ so } \Delta H = \Delta E + \Delta(PV)$$

The pressure is constant, so $\Delta(PV) = P\Delta V = P(V_{gas} - V_{liquid})$. The volume of 1 mole of liquid water is approximately 18 mL (18.0 g divided by the density of water, about 1 g/mL). The volume of a mole of water vapour is about 31.01 L at 373 K, from the ideal gas law. Clearly the liquid volume is neglible. It follows that $\Delta(PV) = PV_{gas} = 31.01$ L-atm. Conversion to joules is accomplished using the ratio of the gas constant in joules to the gas constant in L-atm:

31.01 atm-L * (8.314 J/0.08206 L-atm) = 3140 J = 3.14 kJ

Solving equation for ΔE gives:

$$\Delta E = \Delta H - PV_{gas} = 40.67 - 3.1 = 37.6 \text{ kJ}$$

DH and DE have similar values. The small difference of 3.1 kJ is a measure of the work done on the surroundings during the vaporization process, as the gas pushes back the atmosphere. This calculation substantiates our earlier decision to ignore the usually small difference between DE and DH.

Example: ΔH_{vap} of ethyl alcohol at its normal boiling point,

78°C, is 38.58 kJ/mole. Calculate the enthalpy change when 25.0 g ethyl alcohol vapour condenses at 78°C.

Solution: First note that the stated enthalpy is not a standard value because it applies at 78, not 25°C. It therefore has no superscript o appended. Condensation is the reverse of vaporization. Since vaporization is endothermic, condensation is exothermic, and

$$\Delta H_{cond} = -\Delta H_{vap}.$$

The desired enthalpy change can be calculated from the number of moles of ethyl alcohol and ΔH_{cond}:

moles alcohol = 25.0g/(46.069 g/mole) = 0.543 moles

DH for process = 0.543 moles x -38.58 kJ/mole = –20.94 kJ

20.94 kJ of heat will be released to the surroundings when the vapour condenses.

We have discussed phase changes carried out by heating at constant pressure. We have referred to these as temperature-induced phase changes. We now discuss the energetics of pressure-induced phase changes. These are accomplished by varying the pressure exerted on the substance while keeping its temperature constant. Consider a sample of a substance in the gas phase, confined in a cylinder/piston apparatus as in Figure.

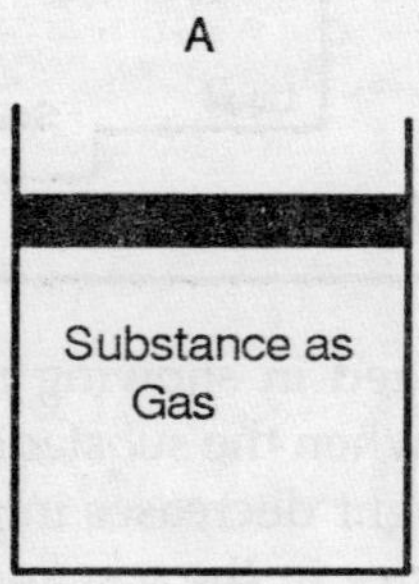

Fig. Pressure-Induced Phase Change

The temperature of the cylinder and the gas is maintained constant at T_1. While keeping the temperature fixed, we gradually increase the pressure on the gas by pushing on the piston. This causes the molecules to crowd more closely together. However, it does not affect the average speeds of the molecules. They will collide with other molecules more frequently and change speed more often, but as long as temperature does not change, the average kinetic energy will not either. When pressure gets high enough, the gas will liquefy, and the potential energy of the molecules will correspondingly decrease.

This energy will be released as heat, and the molecules will temporarily speed up. However, as this heat is transferred to the constant temperature bath through the walls of the cylinder, the gas temperature will return to T_1 and the average molecular speed will return to the original value. Further increase in pressure, still maintaining T constant, causes the liquid to solidify. Again there will be a temporary increase in molecular speed when the molecular potential energy decreases, but this will return to the original value as the heat of fusion is transferred to the constant temperature surroundings. A plot of the potential energy (enthalpy) of the system as a function of applied pressure is shown in Figure.

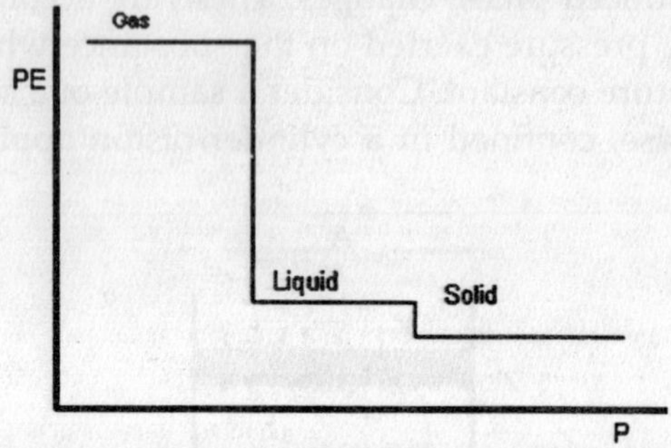

The plot is idealized in showing no change in potential energy with pressure when the substance is present as a single phase. In actuality, slight decreases in potential energy occur with increasing pressure in the gas and liquid phases, and a slight increase occurs in the solid phase. However, these changes are small compared with the changes that occur during phase transformation. As long as temperature remains constant, the kinetic energy of an average molecule remains the same at 3kT/2, regardless of phase, and the speeds obey the Maxwell-Boltzmann Distribution.

The thought experiment described above can be done in

the laboratory using carbon dioxide, CO_2. Starting with a sample of CO_2 gas at -50 °C and 1 atm pressure, an increase of pressure to about 5.3 atm causes liquefaction at a constant temperature of -50 °C. A further increase in pressure to about 20 atm causes the liquid to solidify at -50 °C.

Application of Hess's Law to the process of conversion of solid to gas directly (path 1) or indirectly via the liquid phase (path 2) leads to the conclusion that for any pure substance

$$\Delta H_{sub} = \Delta H_{fus} + \Delta H_{vap}$$

This is indicated in terms of the potential well in Figure.

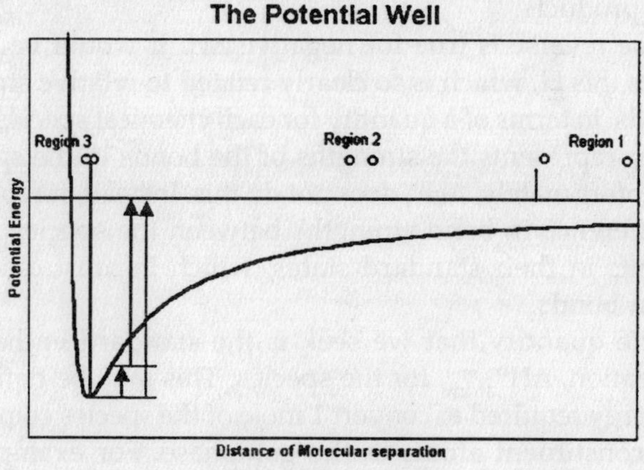

Fig. Potential Well

ENTHALPY OF ATOMIZATION

Tabulations of standard thermodynamic data for chemical compounds are invariably based on the standard enthalpy of formation, $\Delta H^o{}_f$, discussed earlier.

We have seen that this quantity is useful for calculating the overall enthalpy changes for a huge number of chemical reactions.

In this section, we will explore an alternative approach to the tabulation of thermodynamic enthalpy data, one that more directly emphasizes the importance of chemical bonding in chemical reactions.

Spontaneous conversion of a particular set of reactants to a particular set of products either does or does not tend to occur. As we will learn, there are two major factors involved in whether a reaction occurs. One of these is an energy factor, which we can summarize as follows:

A reaction tends to occur if the bonds between atoms in the products are stronger than the bonds between atoms in the reactants.

The enthalpy change measures directly the relative strengths of bonds in reactants and products. If ΔH is positive, we interpret this to mean that more energy is required to break the bonds in the reactants than is released in forming the bonds of the products.

The reverse is true for negative ΔH. It would be nice to express this H, which is so clearly related to relative strengths of bonds, in terms of a quantity for each chemical species which directly represents the strengths of the bonds in the species.

Unfortunately, ΔH^o_f does not do this. Instead, it represents the difference in bond strengths between the species and its elements in their standard states, which in most cases also contain bonds.

The quantity that we seek is the standard enthalpy of atomization, ΔH^o_{atom}, for the species. This may be defined as the energy required to convert 1 mole of the species completely to its constituent atoms in the gas phase. For example, the standard enthalpy of atomization of $CaCO_3$ is the enthalpy change for the following process:

$$CaCO_3(s) \rightarrow Ca(g) + C(g) + 3\ O(g)\ [\Delta H^o_{atom}]$$

There are no chemical bonds between the isolated atoms in the gas phase. Thus ΔH^o_{atom} represents the total enthalpy change required to break all of the bonds in 1 mole of calcium carbonate. It directly measures the strengths of the bonds in this material. Breaking chemical bonds is always endothermic; that is, it always requires the input of energy:

$$\text{bonds} \rightarrow \text{no bonds}\ [\Delta H > 0]$$

Consequently, enthalpies of atomization are invariably positive numbers. Let's apply this idea to the general chemical reaction in equation,

$$R \rightarrow P\ [\Delta H]$$

Hess's Law to this process by first converting the reactants to elements in their standard states. These were then allowed to form products in their standard states.

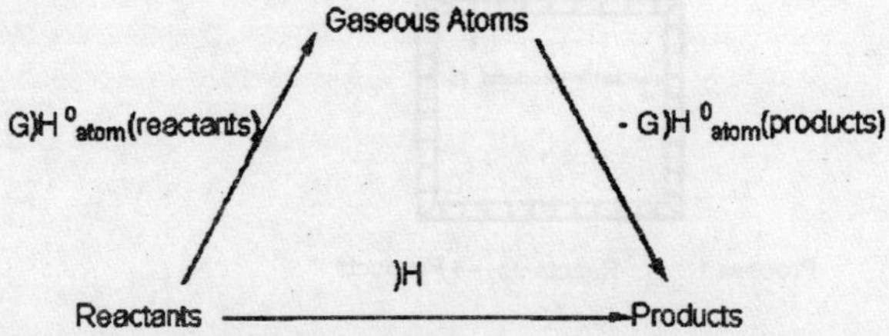

Fig. Alternate Hess's Law Path

Here the chemical and physical bonds in the reactants are broken by the input of an amount of energy equal to the total of the atomization enthalpies of all of the reactants. This converts the reactants to atoms, in the gas phase. These atoms are then allowed to come together in the configurations corresponding to the products.

The bonds that form in this process release an amount of energy equal to the negative of the total atomization enthalpy of the products. The enthalpy of reaction is thus seen as the net result of breaking the reactant bonds and replacing them with product bonds.

Since reaction is perfectly general, we may write,

$$\Delta H^o_R = \Sigma[\Delta H^o_{atom}(\text{react})] - \Sigma[\Delta H^o_{atom}(\text{prod})]$$

The importance of bonds in determining the energy change of a reaction is thus directly seen. Table contains standard enthalpies of atomization for a number of elements and compounds.

EXPERIMENTAL MEASUREMENT

The measurement of enthalpy changes of chemical and physical processes is accomplished by calorimetry. The chemical or physical process is carried out on a precisely known scale in an insulated container called a calorimeter.

The insulation prevents the flow of heat from the surroundings to the system (inside the calorimeter) or vice versa. The accompanying diagram summarizes the calorimetric experiment pictorially.

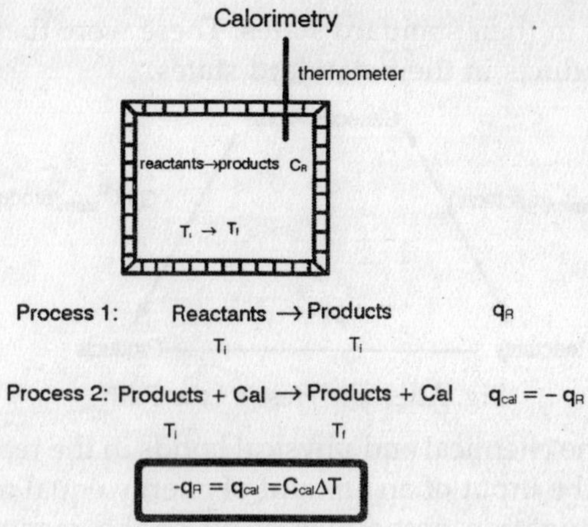

Fig: Calorimetry

Inside the calorimeter, the heat produced or consumed by the process is absorbed or provided by the calorimeter, which will cause the temperature of the calorimeter and its contents to change.

From the measured temperature change and the known heat capacity of the calorimeter, the heat absorbed or provided by the calorimeter can be calculated using equation 9-4 ($q_{cal} = C_{cal}\Delta T$).

For an exothermic process, for which $q_{process} < 0$, the heat produced is absorbed by the calorimeter. Therefore q_{cal} is > 0 and the temperature increases. For an endothermic process ($q_{process} > 0$), $q_{cal} < 0$ and temperature decreases. In either case, $q_{cal} = -q_{process}$. Substituting $-q_{process}$ for q_{cal} gives the fundamental equation of calorimetry: $-q_{process} = C_{cal}\Delta T$ $q_{process}$ can then be adjusted for the scale of the reaction to give a value per mole of process, $q_{process,m}$. If the process is carried out at constant pressure, $q_{process,m} = \Delta H_{process}$ (approximately = $DE_{process}$); if at constant volume, $q_{process,m} = \Delta E_{process}$ (approximately = $DH_{process}$).

Example: 1.765 g ethyl alcohol is burned in excess oxygen in a constant volume calorimeter. The temperature of the calorimeter rises from 298.00 K to 299.51 K. C_{cal} = 34.65 kJ/K. What is the standard heat of reaction for combustion of ethyl alcohol?

Solution: Apply equation:

$q_{process} = -C_{cal}DT = -(34.65 \text{ kJ/K})(299.51 - 298.00)$
$= -52.32 \text{ kJ}$

This is the heat of reaction for burning 1.765 g ethyl alcohol. We can convert this to a standard heat (a heat per mole) by dividing by the number of moles of ethyl alcohol.

$q_{process,m} = -52.32 \text{ kJ}/(1.765 \text{ g}/46.069 \text{ gmole}^{-1}) = -1366 \text{ kJ}$

Since the reaction was done at constant volume, $q = DE_R^o = -1366$ kJ. As we expect by now, DH_R^o is very close to this (-1368 kJ).

Example: A series of experiments was carried out to determine the relationship between the enthalpies of reaction of hydrochloric acid, HCl, and sulfuric acid, H_2SO_4, with the strong base, sodium hydroxide, NaOH. The expected reactions are

$HCl + NaOH \rightarrow NaCl(aq) + H_2O$

$H_2SO_4(aq) + 2NaOH \rightarrow Na_2SO_4(aq) + H_2O$

It was anticipated that the enthalpy of reaction for sulfuric acid should be twice that for hydrochloric acid, because the former contains two hydrogen ions, while the former contains only one.

The experiments were carried out as follows. First, the reaction of HCl with NaOH was investigated using 2.0 M solutions of acid and base. Several volumes of acid solution (between 2 and 40 mL) were reacted with the same volume of base solution (50 mL). In all cases, the total volume of solution in the calorimeter was adjusted to 100 mL by addition of the necessary volume of water (the difference between 50 mL and the volume of acid to be used) to the calorimeter along with the acid solution.

The acid solution in the calorimeter and the base solution outside the calorimeter were adjusted to the same temperature, and the base was quickly added to the calorimeter. The resulting temperature change was measured. Data are presented in Table 9-5; observed temperature change is plotted against the millimoles of acid used (volume used multiplied by molarity of the acid solution). The same procedure was then followed using 2.0 M H_2SO_4 and 2.0 M NaOH. The results of these experiments are also shown in the table and figure.

The heat capacity of the calorimeter containing 100 mL of aqueous solution is known to be 350 J/°C. What is the relationship between the enthalpies of reaction of the two acids with NaOH? What are the values of the reaction enthalpies?

Discussion. Several conclusions can be drawn from the figure. First, for both acids, the observed temperature change is directly proportional to the moles of acid used. Because the acid is the limiting reagent, this is equivalent to saying that the temperature change is proportional to the amount of reaction. This substantiates that reaction enthalpy is an extensive property—its value depends on the amounts of substances reacted. The slopes of the two linear plots give the temperature change per millimole of acid.

The values obtained from the best linear fit of the data are,

$$\Delta T/\text{mole HCl} = 0.160\ ^{\circ}\text{C/mmole}$$
$$\Delta T/\text{mole } H_2SO_4 = 0.385\ ^{\circ}\text{C/mmole}$$

Second, it is clear that the value for sulfuric acid is somewhat more than twice as large as the HCl value. The ratio of slopes is 2.41. There is minimal scatter around the best fit straight lines. It is thus hard to justify rounding 2.41 to 2.00, the anticipated value. The enthalpy of reaction of H_2SO_4 with NaOH is about 20% larger than expected. This requires explanation.

Strong acids such as HCl react with water as follows:

$$HCl + H_2O \rightarrow H_3O^+ + Cl^-$$

Reaction of such an acid with a base such as NaOH actually involves the following reaction:

$$H_3O^+ + OH^- \rightarrow 2H_2O$$

In anticipating that the enthalpy of reaction for sulfuric acid would be twice that for hydrochloric acid, we assumed that sulfuric acid readily gives up both hydrogen ions to water:

$$H_2SO_4 + 2\,H_2O \rightarrow 2\,H_3O^+ + SO_4^{2-}$$

If this were true, however, the temperature change per mole of sulfuric acid would be exactly twice the value for hydrochloric acid. The fact that it is not reveals to us that sulfuric acid releases the first hydrogen ion readily, but not the second one. When dissolved in water, sulfuric acid releases the first hydrogen ion as follows:

$$H_2SO_4 + H_2O \rightarrow H_3O^+ + HSO_4^-$$

The second hydrogen ion, however, remains attached to the sulfate anion. When base is added, H_3O^+ produced by the first hydrogen ion reacts with OH- as expected. In order for the second proton to react with OH-, it must be removed from the sulfate ion:

$$HSO_4- + OH- \rightarrow H_2O + SO_4^{2-}$$

Hess's law allows us to formulate this in terms of the following two steps:

$$HSO_4^- + H_2O \rightarrow H_3O^+ + SO_4^{2-} \; \Delta H_1$$

$$H_3O^+ + OH^- \rightarrow 2H_2O \; \Delta H_2$$

The extra 20% of heat produced in the sulfuric acid-sodium hydroxide reaction is ΔH_1.

Finally, the enthalpy per mole of reaction can be obtained by multiplying the temperature change per mole by the heat capacity of the calorimeter:

ΔH_K for HCl = –0.160 °C/mmole * 350 J/°C = –56.0 J/mmole = –56.0 kJ/mole

ΔH_R for H_2SO_4 = –0.385 °C/mmole * 350 J/°C = –134.8 J/mmole = –134.8 kJ/mole

From these values, we obtain ΔH_1 = –22.8 kJ/mole of HSO_4^-

BOND ENERGY

Structures that result from covalent bonding between atoms. There we developed methods for understanding and predicting the number of bonds in a molecule or polyatomic ion, and the three-dimensional distribution of atoms in space that results from the bonding.

In this chapter, we have stressed that chemical reactions involve bonding rearrangements, and that the energy changes that accompany reactions reflect the relative strengths of bonds in reactants and products. In this section, we discuss the relationship between the enthalpy change for a reaction and the strengths of specific covalent bonds in reactants and products.

The total bond energy of a gaseous molecule is the energy required to decompose the molecule to isolated, gaseous atoms. For example, the bond energy for $CO_2(g)$ is the enthalpy change for the following reaction:

$$CO_2(g) \rightarrow C(g) + 2\ O(g)\ \Delta H = 1609\ kJ = \text{total bond energy (BE)}$$

This is simply the enthalpy of atomization of $CO_2(g)$. We would now like to partition this total enthalpy among the individual bonds in the CO_2 molecule. The rupture of two moles of C=O bonds in CO_2 to form gaseous atoms requires 1609 kJ of energy; thus the energy required to break one C=O double bond is 804.5 kJ/mole. Similarly, one fourth of the total bond energy (enthalpy of atomization) of methane, CH_4, is the average strength of a C-H single bond in methane.

Total bond energies for a great many molecules have been measured. The results indicate that the strength of, say, an I-F bond is very much (although not exactly) the same no matter what molecule it occurs in.

The data below support this statement:

Compound	**I-F Bond Strength**
IF	278
IF3	272
IF5	268
IF7	231

It has thus been possible to assign an average bond energy to each common bond type. Table 9-6 summarizes bond energy data for a large number of chemical bonds. Note that all of the entries in the table are positive numbers; bond breaking is always an endothermic process.

Using the bond energy information in the table, it is possible to estimate the value of ΔH for a large number of gas phase chemical reactions, using the simple Hess's Law cycle. It is important to stress that this is an estimate, because the tabulated bond energies are average values. According to the figure, we visualize a pathway for the reaction in which reactants are first converted to gas phase atoms, then the atoms are allowed to recombine to form products. Disassembling the reactants requires the input of an energy equal to the total energy of all of the bonds in the reactants.

This quantity, Σ[BE(reactants)] is invariably positive. Assembling the products releases the total bond energy of the products, Σ[BE(products)]. This step is exothermic, as bonds are formed. If the total bond energy of reactants exceeds that of products, the reaction is endothermic (i.e., more energy must be put in to break the reactant bonds than is produced in forming product bonds); if the bond energy of products exceeds that of reactants, the reaction is exothermic. The value of ΔH for the reaction may thus be estimated using equation: $\Delta H^o_R = \Sigma$ [BE(reactants) – Σ BE(products)]

Example: Using bond energy data from Table 9-6, estimate ΔH for the following reaction:

$$C_2H_2(g) + 5/2\ O_2(g) \rightarrow 2CO_2(g) + H_2O(g)$$

Solution: We first tabulate the bonds present in reactants and products, then use Table 9-6 to assign strengths to these bonds:

Reactants	**BE**	**Products**	**BE**
C(3)C	835	C=O (4)	4 × 799
C-H (2)	2 × 411	O-H (2)	2 × 459
O=O (2.5)	2.5 × 494		
total:	2892	total:	4114

Bonds in the products are stronger than in the reactants. The reaction is exothermic in the amount 1222 kJ. Note that this value is similar to, but not exactly the same as, the value calculated using standard enthalpies of formation (you should work this out). The standard enthalpy of formation is a very accurate quantity, which applies specifically to a particular compound. In contrast, bond energies are average values, and thus cannot be expected to reproduce exactly the specific bond strengths in a particular compound. Enthalpy changes calculated using average bond energies are approximations.

CHEMICAL REACTIONS

Exothermic processes produce heat, and in the process

adopt a final atomic arrangement that has lower potential energy (enthalpy) than the initial arrangement. Many such processes are spontaneous — that is, they occur naturally, without assistance from outside the system. For example, the rusting of iron is a spontaneous process that releases 825 kJ of heat per mole of iron(III) oxide produced; natural gas spontaneously burns in oxygen, releasing 890 kJ per mole of CH_4; and hydrogen gas reacts spontaneously with oxygen gas to form water, with the release of 286 kJ of energy per mole of water in the form of heat and some light. It was at one time thought that all spontaneous chemical processes were exothermic, based on the well-known mechanical tendency to minimize potential energy (e.g., water flows downhill).

Thus the drive to decrease the potential energy of the system was thought to be the driving force for chemical reactions. In fact, although many spontaneous chemical processes are exothermic, some are endothermic. For example, ice melts spontaneously at 25°C, even though the process is endothermic. Ammonium nitrate, NH_4NO_3, spontaneously dissolves in water, despite the fact that a substantial quantity of heat is transferred to the system from the surroundings in the process (the water becomes cold as the salt dissolves).

The drive to minimize potential energy (enthalpy) is clearly an important factor in determining spontaneity, because so many spontaneous processes are exothermic. However, it is not the only factor.

SECOND LAW OF THERMODYNAMICS

We introduced the concept of disorder by discussing the free expansion of an ideal gas. Common experience tells us that this process is spontaneous—it occurs without outside assistance.

The change in potential energy for this process is zero, so its spontaneity is not due to the tendency to minimize PE. Instead, the process is driven by the tendency to increase disorder. Our earlier discussion of disorder was vague in two respects. First, we did not distinguish between the disorder of the system and that of the surroundings. Second, disorder was treated only qualitatively. We will put the disorder concept on a much firmer basis by making it quantitative.

The free expansion of an ideal gas, expansion of the gas (the system) increases its disorder because each gas molecule has a larger volume to explore. Disorder is greater because knowledge of the location of a gas molecule is less. Clearly, then, free expansion leads to an increase in disorder—in entropy—of the system: $\Delta S_{sys} > 0$. What happens in the surroundings when the gas expands? Other than the small effort expended in opening the connecting valve, absolutely nothing happens. No boundary of the system moves, so the surroundings does not experience work. There is no heat flow to or from the surroundings because the gas is ideal and thus does not change internal energy on expansion. Absolutely nothing is experienced by an observer in the surroundings. It therefore follows that the disorder in the surroundings is unchanged by the free expansion of the gas: $DS_{surr} = 0$. The system and surroundings together constitute the universe. Combining our two results gives equation for the free expansion of an ideal gas:

$$\Delta S_{sys} + \Delta S_{surr} = \Delta S_{univ} > 0$$

Stated in words, equation says that in any spontaneous process, the entropy of the universe increases. Although the equation was developed by consideration of a single simple process, it has proved to have far-reaching and profound generality: it is true for any spontaneous process. Thus for the burning of natural gas; the rusting of iron; the respiration of living organisms; and the light-promoted process of photosynthesis, the disorder of the universe increases. The simple and powerful statement in Equation is called the Second Law of Thermodynamics.

The Second Law is in many respects more subtle and difficult to understand than is the First Law. Yet it is the Second Law that allows prediction of spontaneity, so to gain some understanding of it is crucially important. To facilitate understanding, we will make liberal use of examples the important ideas; and we will set down a number of simple guidelines for predicting the sign of the entropy changes for system and surroundings in physical and chemical processes. These guidelines are not hard and fast rules—instead, they give the correct result most, but not all, of the time. However, their simplicity makes them worth remembering.

The search for a quantitative measure of entropy was a long and difficult one. The analysis of steam engines (the original genesis of thermodynamics) was the source of one approach, formulated by Rudolf Clausius in the 1880s. The Clausius formulation expresses entropy in terms of heat and temperature. The second approach, that of Ludwig Boltzmann, was based on considerations of probability and statistics. Although the two quantitative formulations of entropy appear different, it has been shown that they are equivalent in any application to which they may both be applied. We will briefly consider these two approaches to entropy, beginning with the Clausius formulation.

ENTROPY AND HEAT

That there is a connection between disorder and heat is somewhat expected. First, heat, the random thermal motion of molecules, is clearly the most disordered and non-useful form of energy. Second, much of our experience indicates that disorder is produced when heat is added to systems. The solid form of a substance (water, sodium chloride, iron) melts, and the liquid form boils, when heat is added. Gases expand when heated. Atoms form a plasma of nuclei and electrons when heated to the temperatures of stars. These qualitative realizations help us to appreciate the Clausius definition of entropy in equation.

$$\Delta S_{sys} = q_{max}/T$$

Stated in words, the equation says that the change in entropy in a system when a process is carried out may be calculated by dividing the amount of heat added to the system by the temperature of the system. Strictly speaking, equation is valid only when the temperature of the system remains constant. Most of the processes that we will consider occur at constant temperature, however, and equation applies.

The subscript "max" on the heat term represents a

restriction on the equation. It indicates that when calculating the entropy change for a change in the system from a particular initial to a particular final state, we must use the heat that would be added to the system along the maximum heat path. This is known as the reversible path. Our intention here is merely to gain appreciation for the connection of entropy to heat. Consequently, we will not discuss this idea of maximum heat further here.

We now proceed to examine the "sense" of equation. First, the Clausius formulation of entropy recognizes the clear connection between disorder and heat. The more heat we add, the more disorder we produce, at a given temperature. Second, the equation says that the disorder produced by addition of a given amount of heat is less when the system is at high temperature than when it is at low temperature. This is also reasonable. In a system at low T (say, near absolute zero) the kinetic energy of random molecular motion is small. At high T (say 1000 K), however, the thermal energy is substantial.

Addition of a specified amount of heat to such a system produces a much larger fractional change in disorder at low T than at high T. Equation recognizes that if a system is already highly disordered, a small amount of heat will not make much difference. For example, when your room has not been cleaned for a month (or longer), the addition of one more dirty sock to the mess will hardly be noticeable.

In contrast, if your room has just been cleaned (lowered in entropy), that dirty sock on the floor will be very noticeable. Third, the equation tells us that the units of disorder are those of energy divided by temperature, or J/K. Finally, entropy is represented in equation by an upper case symbol, which we reserve for state functions. Entropy is a state function. This means that the change in entropy of a system is independent of the path taken between initial and final states. Two simple and useful guidelines can be distilled from equation. Both guidelines refer to systems consisting of a single pure substance.

Guideline: If heat is added to a pure substance, the entropy of the substance must increase.

Guideline: If heat is removed from a pure substance, the entropy of the substance must decrease.

Example: In which member of each pair is entropy (disorder) higher?

Fe(s) at 0 °C; Fe(s) at 25 °C

H_2O(l) at 298 K; H_2O(g) at 398 K

Ar(g) at 300 K; Ar(g) at 150 K

Solution: To convert Fe(s) at 0 °C to Fe(s) at 25 °C requires the addition of heat. As a result of the addition of heat, the average kinetic energy of the iron atoms increases, with a corresponding increase in disorder in the system. Fe(s) has higher entropy at 25 ° than at 0 °C.

To convert H_2O(l) at 298 K to H_2O(g) at 398 K requires the addition of heat. First, liquid water must be heated from 298 K to the boiling point, 373 K. This increases the thermal motions, and the disorder, of the molecules of the liquid. Second, liquid water must be boiled at 373 K. This frees water molecules from the potential well of the liquid and allows them to move independently of one another; disorder increases substantially. Finally, water vapour at 373 K must be heated to 398 K. Again, average molecular kinetic energy (and disorder) increases with the addition of heat. H_2O(g) at 398 K has substantially higher entropy than H_2O(l) at 298 K.

To convert Ar(g) at 150 K to Ar(g) at 300 K requires the addition of heat. This increases the random thermal molecular motion, and the disorder of the system. Ar(g) has higher entropy at 300 K than at 150 K.

BOLTZMANN INTERPRETATION OF ENTROPY

The Clausius interpretation of entropy is a pragmatic one, based on engine analyisis. Ludwig Boltzmann took a different aproach, based on the idea that there is a parallel between the amount of disorder in a system in a particular state and the number of ways that the system may be arranged in that state. In a very general way, this can be stated as

$$S = f(W)$$

where W = the number of arrangements possible for a particular state of a system, and f is a to-be-determined function. We suppose at the outset that a state with only one possible arrangement has no disorder—that is, $S = 0$—and that S should increase smoothly as W increases. Thus our first constraint on the form of the function, f, is that $f(1) = 0$. To arrive at a second constraint on f, consider a simple example. Shown is a row of 10 buckets placed side-by-side.

A ping pong ball is tossed at the row of buckets and lands in one of them (we assume for simplicity that there are no misses—the ball always goes in one of the buckets). The state of the resulting system is described by saying that there is one ball in the set of 10 buckets. Because the ball can be in any of the buckets, the number of arrangements for the state, W, is 10. There is thus some disorder associated with this state. An observer asked to view the row of buckets containing one ball would be uncertain as to where the ball is. His uncertainty is a measure of the disorder.

We now double the amount of material in the system by tossing a second ball, which also lands in one of the buckets. The state of the system is described by saying that there are 2 balls in the set of 10 buckets. There are 10 possible locations for the first ball. For each of these possibilities, there are 10 possible locations for the second ball. The total number of arrangements for the state is $10 \times 10 = 10^2$. By similar reasoning, if n balls in succession were tossed, the number of arrangements would be 10^n. In general, for a state consisting of N balls placed in L locations, the number of arrangements is given by equation.

$$W = L^N$$

The entropy of the state is $S = f(L^N)$. We now impose the second constraint on the form of the function f: S must double when the amount of material in the system (N) doubles. In equation form,

$$f(L^{2N}) = 2f(L^N)$$

The function that satisfies both imposed constraints ($f(1) = 0$ and $f(L^{2N}) = 2f(L^N)$) is the logarithm. Thus

$$S \text{ proportional to } \ln W = C \ln L^N = NC \ln L$$

where C is a proportionality constant. Equation is the Boltzmann interpretation of entropy. According to equation, the entropy of a system is directly proportional to the number of particles (such as ping pong balls, or molecules), N, in the system and to the natural logarithm of the number of locations each particle can have. Thus the amount of disorder goes up directly with the amount of matter in the system, in agreement with the second constraint. When each particle has only one allowed location, L is 1 and S is 0. Further, we see that as the number of allowed locations for a particle (L) goes up, S increases more and more slowly.

This is seen below:

L	S	Increment in S
1	0	
2	.693 NC	.693 NC
3	1.10 NC	.41 NC
4	1.39 NC	.29 NC
5	1.61 NC	.22 NC
.		
.		
99	4.60 NC	
100	4.61 NC	0.01 NC

When the system has low entropy initially (L is small, say 1), the addition of one more allowable location has a dramatic effect on S. Thus increasing L from 1 to 2 increases S by 0.693 NC. However, when the system has high entropy initially (L is large, say 99), increasing L by 1 increases S by only 0.01 NC, a negligible effect.

Boltzmann expressed equation in somewhat different form, $S = k \ln W$

The constant, k, is called Boltzmann's constant. It has the value 1.38×10^{-23} J/K. A plot of S versus W, the number of arrangements of N particles in L locations. The slope of the plot is steep when L is small, but approaches zero when L becomes large, showing that DS becomes smaller for a given value of DW as S (and W) increases.

Thus when W increases from 1 to 200, S increases by about 7×10^{-23} J/K; but when W increases by the same amount from 800 to 1000, S increases by only 0.3×10^{-23} J/K. The following mathematical operations, based on equation, show analytically what Figure shows visually.

Because $W = L^N$, where N = the number of particles in the system and L = the number of options per particle,

$$S = k \ln W = k \ln L^N = Nk \ln L$$

Doubling N gives $S = k \ln L^{2N} = 2Nk \ln L$. Thus S goes up in direct proportion to N. However, doubling L gives $S = k \ln (2L)^N = Nk (\ln 2 + \ln L) = Nk \ln L + Nk \ln 2$. In this case, S goes up incrementally by N ln 2 each time the number of options per particle increases by a factor of 2.

The Boltzmann and Clausius interpretations of entropy are thus in agreement that the more disordered a system is initially, the less difference an incremental change in the number of arrangements makes.

If a state, "a", has more possible arrangements than a state, "b", then state "a" is more probable than state "b". The Boltzmann interpretation of entropy explicitly recognizes the connection between entropy and probability that we developed intuitively. According to this interpretation, the second law of thermodynamics says that the universe seeks states of higher probability.

Example: Consider a system consisting of a pair of dice in a closed box. Suppose that we have no information about this state of the system other than that there are 2 dice in a box. How many "particles" constitute this system? How many arrangements are possible for the system in this state? What is the entropy of the system in this state?

Now suppose that the system is known to be in a "state" in which the sum of the numbers showing on the top faces of the two dice is 2. What is the entropy of this state? What is the entropy of the state in which the number total is 7? The most probable state of the system is the one for which we know only that there are two dice in a box. What is the next most probable state (numerical total) of the system? Discuss the entropies of the least and most probable numerical totals (states).

Solution: The first state to be considered is the one described by saying that there are 2 dice in a box. We will consider each die to be a particle. Then the system consists of

2 particles. Each particle can be in any of 6 "locations" (a die can have 1, 2, 3, 4, 5, or 6 showing on its top face). The total number of arrangements of the 2 particles in the described state is $6 \times 6 = 6^2$, or 36. This is the number of arrangements possible for the system so

$$S = k \ln W = 1.38 \times 10^{-23}\ \text{J/K} * \ln (6^2) = 4.95 \times 10^{-23}\ \text{J/K}$$

To assess the probabilities and entropies of the other described states (numerical totals of 2 and 7), we must figure out the number of ways to obtain each state:

State (Numerical Total)	Die 1	Die 2	# Arrangements
2	1	1	1
3	1	2	
	2	1	2
4	1	3	
	2	2	
	3	1	3
5	1	4	
	2	3	
	3	2	
	4	1	4
6	1	5	
	2	4	
	3	3	
	4	2	
	5	1	5
7	1	6	
	2	5	
	3	4	
	3	4	
	5	2	
	6	1	6
and so on.			

The most probable state is 7, because 6 different arrangements of the system produce it. The least probable states are 2 and 12, because each is produced by only one arrangement. State 7 has highest probability and highest entropy, with value k ln 6; states 2 and 12 have lowest probability and lowest entropy, with value k ln 1 = 0.

The entropy associated with states (numerical totals) of dice may be thought of as follows. A blindfolded person is told that the dice show a total of 2, and is asked what number is showing on each die. S/he can, with certainty, state that each die shows 1. In contrast, when told that the dice show a total of 7, s/he can only guess, with 1 chance in 6 of being correct, at the numbers on individual die. S/he is uncertain. Finally, when told only that there are 2 dice in a box, s/he has only 1 chance in 36 of guessing the numbers on individual die. The extent of uncertainty is a measure of the entropy of the state of the system.

ENTROPY CHANGES

We now turn to the problem of entropy changes in chemical reactions. To calculate these will require that we have available quantitative data on the entropy of 1 mole of each pure substance participating in the reaction. Such data are obtained experimentally by judicious combination of the Boltzmann and Clausius interpretations of entropy. Before looking at things quantitatively, though, it is best to develop a qualitative feel for entropy changes in processes.

QUALITATIVE CONSIDERATIONS

Is it possible to predict simply by examining the equation for a physical or chemical process whether or not the process is spontaneous from left to right? The answer: sometimes. In order to make such a prediction, it is necessary to demonstrate that DS_{univ} for the process is greater than zero. This in turn requires consideration of both ΔS_{sys} and ΔS_{surr}:

$$\Delta S_{univ} = \Delta S_{sys} + \Delta S_{surr}$$

If we can demonstrate that $DS_{univ} > 0$ for a process of interest, we may conclude that the process will spontaneously occur. If instead we demonstrate that $\Delta S_{univ} < 0$ for the process, we may conclude that the process is not spontaneous, but that the reverse process is. It is important to realise that DS_{univ} may be positive even when either ΔS_{sys} or ΔS_{surr} (but not both) is < 0. Only the total of the two contributions matters. Thus entropy may decrease in the system by, say, 100 J/K, as long as it increases in the surroundings by more than 100 J/K, so that ΔS_{univ} is > 0. Similarly, S may decrease in the surroundings, as long as it increases more in the system. We have already had quite a lot to say about the entropy of systems consisting of single pure substances, S_{sys}, and the factors affecting it. We must now extend these ideas to processes in which more than one pure substance may be involved. Several guidelines are useful:

Guideline: ΔS_{sys} for a chemical or physical process is positive when

- The number of moles of gas increases;
- The number of independent particles increases;
- The system volume increases at constant temperature;
- Bonds are broken.

Example: Predict the sign of ΔS_{sys} for each process, and explain the prediction.

$$CH_4(g) \rightarrow C(g) + 4\,H(g)$$
$$H_2O(g) \rightarrow H_2O(l)$$

Solution: In the process,

$$CH_4(g) \rightarrow C(g) + 4H(g),$$

bonds are broken and no bonds are formed to replace them. Disorder will surely increase in this process, and $\Delta S_{sys} > 0$. Note that both the number of independent particles and the number of moles of gas also increase, reinforcing our prediction. Because bonds are broken in this process, we also predict $\Delta H > 0$; bond breaking is always endothermic. Because this process involves atomization of 1 mole of methane, ΔH for the process is the enthalpy of atomization of methane.

In the process $H_2O(g) \rightarrow H_2O(l)$, a highly disordered gas is converted to a relatively organized liquid. Disorder decreases, and $\Delta S_{sys} < 0$. We expect condensation of vapour to be exothermic: $\Delta H < 0$. Energy leaves the system and enters the surroundings in this process. From the two processes considered in Example, we can develop an important and useful generalization. For a process in which the bonds of a substance are ruptured, both ΔS and ΔH are > 0. When bonds are broken, the system climbs out of a potential energy well.

This requires that molecular forces be overcome, a process that is always endothermic: $\Delta H > 0$. When forces are overcome, increased particle freedom inevitably results: $DS > 0$.

$$\text{Bonds} \rightarrow \text{no bonds } \Delta H > 0; \Delta S > 0$$

$$\text{Aggregated state} \rightarrow \text{deaggregated state } \Delta H > 0; \Delta S > 0$$

When methane is atomized—that is, deaggregated—covalent bonds are ruptured, and the five atoms are free to move independently; when water condenses (aggregates), bonds are formed as intermolecular forces operate in the liquid phase. The freedom of motion of the molecules is more restricted, and their potential energy is lower than in the gas phase.

Example: Predict qualitatively the sign of ΔS_{sys} and ΔH_{sys} for the reaction,

$$F_2(g) + H_2(g) \rightarrow 2HF(g)$$

Solution: Here we have little help. This can not be viewed as a process that involves only the breaking or making of bonds. Although F-F and H-H bonds must be broken to form fluorine and hydrogen atoms, they are replaced with H-F bonds in the products. The moles of gas stays the same in the process (2 moles in the reactants replaced by 2 moles in the products), as does the number of free particles. It is reasonable in such a case to predict that the entropy change will be small: DS_{sys} approximately 0. However, the enthalpy change depends on the relative bond strengths. We cannot qualitatively predict this without considerable experience. We conclude:

ΔS_{sys} approximately 0

ΔH_{sys} cannot predict

Example conveys another important lesson. In most chemical reactions, bonds in the reactants are broken; but bonds are formed in the products. In the majority of cases, the number of chemical bonds in reactants and products is the same, because the same atoms, with the same valences, are present. In such processes, the vast majority, it is not possible to use equation to qualitatively assess the entropy and enthalpy changes. In most chemical reactions, the atoms proceed from the potential well of the reactants to the potential well of the products. Usually the same number of chemical bonds is involved in both states, and a simple qualitative prediction of ΔS and ΔH based is not possible. Instead, the overall enthalpy change depends upon which of the two potential wells is deeper; and the overall entropy change depends on whether the reactants or the products are more disordered. The process shown in the figure is endothermic; the sign of its entropy change is not discernible from the potential well diagram.

ENTROPY CHANGES IN CHEMICAL REACTIONS

If we know the absolute entropy for each substance in a chemical reaction, it is easy to calculate the entropy change, ΔS^o_R, for the reaction. We find the sum of the absolute entropies of the products and subtract from it the sum of the absolute entropies of the reactants. Of course, if a reaction involves more than one mole of a substance, we must multiply its absolute entropy by this number of moles. The entropy change so obtained is automatically a standard entropy change because the absolute entropies are standard values. That is, the absolute entropy is the entropy of 1 mole of a substance in its standard state at 298 K. In equation form:

$$\Delta\Sigma^o_R = \Sigma[\Sigma^o \text{ (products)}] - \Sigma[\Sigma^o \text{ (reactants)}]$$

Example: Predict the sign of ΔS^o_R for the reaction of Mg(s) with CO_2(g), then calculate its value.

$$2Mg(s) + CO_2(g) \rightarrow 2MgO(s) + C(s)$$

Solution: Our prediction is that the entropy of the solid products will be lower than that of the reactants, one of which is a gas with presumably large Σ^o. Thus ΔS^o_R is expected to be < 0. Now we calculate it, using data from:

$$\begin{aligned}\Delta S^o_R &= \Sigma[S^o(\text{prod})] - \Sigma[S^o(\text{react})]\\ &= (2)(26.94) + 5.74 - (2)(32.68) - 213.74 \text{ J/K}\\ &= -219.48 \text{ J/K}\end{aligned}$$

The system becTed depression is carved out in the centre of the bottom half and filled with magnesium turnings.

The magnesium is then heated with a propane torch until it is hot enough to start to burn. At this point, the top half of the dry ice block is replaced, covering the cup. The reaction begins slowly, but then escalates until the dry ice block glows white from the light emitted by the magnesium as it reacts violently with the CO_2 vapour produced in the cup, and a white smoke of MgO spews forth from the crack between blocks. After watching this spectacular display, we conclude that the reaction is indeed spontaneous, despite the prediction based on the dramatically negative value of ΔS^o_R for the reaction.

The problem with our prediction is that it ignores the entropy change of the surroundings. A Second-Law prediction of spontaneity (or lack thereof) must be based on ΔS for the universe, which includes not only ΔS_{sys}, but also ΔS_{surr}. To rationalize the spontaneity of the Mg-CO_2 reaction, we must calculate ΔS_{surr} in addition to ΔS_{sys}. We can do this with the help of one more guideline:

Guideline: The entropy change for the surroundings is calculated by dividing the heat put into the surroundings by the temperature of the surroundings:

$$\Delta S_{surr} = q_{surr}/T_{surr}$$

But the heat put into the surroundings is the negative of the heat of reaction, q_R.

Thus:

$$\Delta S_{surr} = -q_R/T_{surr}$$

The reaction is done at constant pressure, because it is open to the atmosphere.

Thus:

$$q_R = \Delta H_R. \text{ Finally, then}$$

$$\Delta S_{surr} = -\Delta H_R/T_{surr}$$

The other two substances are elements in their standard states, and therefore have $\Delta H_f^o = 0$.

$$\Delta H^o_R = 2\,\Delta H^o_f(MgO) - 1\,\Delta H^o_f(CO_2)$$
$$= 2(-601.7) - (-393.5)\text{ kJ}$$
$$= -809.9\text{ kJ}$$

Then ΔS_{surr} = 809900 J/298 K = 2718 J/K and ΔS_{univ} = 2718 - 219.5 = 2498 J/K

When the entropy change in its entirety is considered, the reaction is hugely spontaneous.

Standard Entropy of Atomization

The standard entropy of atomization of a substance,

ΔS^o_{atom}, is the entropy change that accompanies the conversion of 1 mole of the substance in its standard state to its constituent atoms in the gas phase at a specified temperature, usually 298 K. Several atomization processes and their corresponding standard entropy changes are shown below:

$$CH_4(g) \rightarrow C(g) + 4H(g)\ \Delta S^o_{atom} = 430.7\ J/K$$

$$H_2O(l) \rightarrow 2H(g) + O(g)\ \Delta S^o_{atom} = 320.6$$

$$C_6H_{12}O_6(s) \rightarrow 6C(g) + 12H(g) + 6O(g)\ DS^o_{atom} =$$

We can make several observations about standard entropies of atomization. First, they are entropy changes, because they characterize processes. Thus they must include the symbol D, in contrast to absolute entropies. Second, they are always positive. Bond breaking always increases disorder, because atoms that were previously bound together can now move independently.

The standard entropy (change) for a chemical reaction of any type can be calculated from the standard entropies of atomization of the reactants and products by imagining a Hess's Law cycle in which, first, the reactants are atomized, requiring the input of the total of the standard entropies of atomization of the reactants; and second, the atoms are recombined to form products, with the release of the total of the standard entropies of atomization of the products. We have used this Hess's Law view of a chemical process several times before, and it should be familiar to you now. The overall change in entropy for the reaction, then, is given by equation:

$$\Delta S_R{}^o = \Sigma[\Delta S^o_{atom}(\text{reactants})] - \Sigma[\Delta S^o_{atom}(\text{products})]$$

Standard entropies of atomization provide an alternative to absolute entropies for the determination of entropy changes in chemical reactions.

GIBBS FUNCTION

Example suggests that, at constant temperature and pressure,

$$\Delta S_{univ} = \Delta S_{sys} - \Delta H_{sys}/T$$

The constant pressure restriction is necessary to use DH_{sys} in place of q; the constant temperature restriction is necessary to use,

$$\Delta S_{surr} = q_{surr}/T,$$

which is an isothermal expression. Equation is interesting because it expresses the entropy change of the universe solely in terms of system functions! To put both sides of the equation into energy units, we rewrite it as,

$$T\Delta S_{univ} = T\Delta S_{sys} - \Delta H_{sys}$$

This suggests that we define a new energy function of the system, usually called the Gibbs Free Energy and symbolized G, such that:

$$G = H - TS$$

A change in G at constant temperature is then given by:

$$\Delta G = \Delta H - T\Delta S$$

and it is clear that ΔG, the change in a system function, measures $T\Delta S_{univ}$. This new function, G, incorporates the influences of both enthalpy and entropy in determining the spontaneity of a process and enables us to reframe the spontaneity criterion of the second law of thermodynamics:

At constant T and P, processes having $\Delta G < 0$ are spontaneous; they tend to proceed left to right as written, favoring the final state. Processes having $\Delta G > 0$ are non-spontaneous (impossible); they tend to proceed right to left as written, favoring the initial state.

The change in Gibbs Free Energy, ΔG, reflects a compromise in the effects of ΔH and ΔS on the spontaneity of a process. For the atomization of methane, bonds are broken in converting reactants to products.

Because bond breaking is always endothermic, yet leads to increased freedom of motion of the atoms involved, both ΔH and ΔS for the process are positive.

Reactants are favored by enthalpy; but products are favored by entropy. Whether the process is spontaneous or not is determined by the relative magnitudes of ΔH and $T\Delta S$ in equation. For the atomization of methane at 298 K, the enthalpy "wins"; the process is not spontaneous from left to right.

Example: Apply the new criterion of spontaneity to the Mg-CO_2 reaction of example.

Solution: We use $\Delta G = \Delta H - T\Delta S$ at 298 K. The enthalpy and entropy changes of reaction have already been calculated in the previous example.

$$\Delta G = \Delta H - T\Delta S = -809.9 - (298)(-0.219) \text{ kJ}$$
$$= -745 \text{ kJ}$$

The reaction is predicted to be spontaneous. Note that some care is required to express the changes in enthalpy and entropy in the same units.

Standard Free Energy of Formation, ΔG_f^o. Application of the free energy criterion to chemical reactions is easy. One approach is to calculate ΔH_R^o and ΔS_R^o from tabulated enthalpies of formation and absolute entropies, then calculate DG_R^o for the reaction using equation.

A second approach is to use so-called standard free energies of formation, which have been tabulated by chemists for many important substances.

The standard free energy (change) of formation of a substance is the free energy change that occurs when 1 mole of the substance is made from its elements in their standard states. The definition is analogous to that of standard enthalpy (change) of formation.

For calcium carbonate, the standard free energy of formation is the free energy change for the following reaction:

$$Ca(s) + C(s) + 3/2\ O_2(g) \rightarrow CaCO_3(s)\ [\Delta G_R^o \text{ is by definition } \Delta G_f^o]$$

Table gives standard free energies of formation at 298 K for calcium carbonate and a number of other common substances. For an element in its standard state, the standard free energy of formation is zero, because there is no change in either enthalpy or entropy when 1 mole of an element is formed from itself. If we apply Hess's Law to the free energy change in the same way, we arrive at equation:

Table. Standard Free Energy of Formation, kJ/mole

Substance	DG_f^o
$CaCO_3$	–1128.76
CaO	–604.17
$CH_4(g)$	–50.8
$C_2H_2(g)$	209.2

$C_2H_4(g)$	68.11
$C_6H_{12}O_6(s)$	–910.4
$CO_2(g)$	–394.4
$H_2O(l)$	–236.81
$MgO(s)$	–569.6

$$DG_R^o = S[DG_f^o(prod)] - S[DG_f^o (react)]$$

We apply this to an important reaction.

Example: Is the photosynthesis reaction spontaneous at 25 °C?

$$6\ CO_2(g) + 6\ H_2O(l) \rightarrow C_6H_{12}O_6(s) + 6\ O_2(g)$$

Solution: To answer the question, we require standard free energies of formation for all participants in the reaction.

Substance	ΔG_f^o, kJ/mole
$CO_2(g)$	–394.4
$H_2O(l)$	–236.8
$C_6H_{12}O_6(s)$	–910.4
$O_2(g)$	0

$$\Delta G_R^o = \Delta G_f^o (C_6H_{12}O_6) - 6\Delta G_f^o (CO_2) - 6\Delta G_f^o (H_2O)$$
$$= -910.4 - 6(-394.4) - 6(-236.8) = 2877\ kJ$$

The reaction is predicted to be non-spontaneous (impossible) at ordinary temperature, yet it occurs on a tremendous scale all around us. An interesting question: why does photosynthesis occur?

Standard Free Energy of Atomization

The standard free energy (change) of atomization of a substance, ΔG^o_{atom}, is the free energy change accompanying the conversion of 1 mole oi the substance in its standard state to atoms in the gas phase.

The standard free energy of atomization is related to the standard enthalpy and entropy changes of atomization via equation, which is just applied to the process of atomization.

$$\Delta G^o_{atom} = \Delta H^o_{atom} - T\Delta S^o_{atom}$$

Tabulated values of ΔG^o_{atom} at T = 298 K. Equation shows how they can be used, according to the usual Hess's Law cycle in which reactants are converted to products via the gas phase atoms, to calculate the overall standard free energy change for any chemical reaction of interest.

$$\Delta G_R^o = S[\Delta G^o_{atom} \text{ (reactants)}] - S[\Delta G^o_{atom} \text{ (products)}]$$

This equation can be used as an alternative to equation involving standard free energies of formation.

Because both the standard enthalpy and entropy of atomization are invariably positive numbers, it is clear from that the standard free energy of atomization can be positive or negative, depending on the relative magnitudes of the two terms. It turns out that the enthalpy term dominates at 298 K for most substances, so most of the values are > 0. At high temperature, the second term becomes dominant, so that many standard free energies of atomization become negative.

The Interpretation of ΔG as "Free" Energy. For a process carried out at constant temperature and pressure, the value of DG provides a measure of the portion of the total energy produced by the process that is "free" to be used for doing work of some kind. This is the origin of the name, free energy. To clarify this, we consider the reaction of hydrogen and oxygen to form water at 298 K.

$$2H_2(g) + O_2(g) \rightarrow 2\ H_2O(l)$$

By now, we are aware that this process is exothermic, and can use standard enthalpies of formation or atomization to show that $\Delta H_R^o = -572$ kJ. It is reasonable to wonder whether all of this released energy can be harnessed to do work; for example, to run an engine. In fact, only a portion of this total released energy can be so used, for reasons that we now consider.

A qualitative assessment of the entropy change for the reaction reveals that $\Delta S_{sys} < 0$. Using absolute entropies, it is a simple matter to show that $\Delta S_R^o = -326$ J/K. For the reaction to be spontaneous, it is necessary that ΔS_{univ} be positive; thus the entropy of the surroundings must increase by a minimum of 326 J/K to guarantee spontaneity. At 298 K, this increase in entropy of the surroundings requires that at least $T\Delta S_{surr} = 97.1$ kJ of heat be provided to the surroundings during the reaction. This heat must be drawn from the total energy produced by the reaction. As a consequence, only 475 kJ of the total 572 kJ is actually available to do work. The remainder must be paid to the surroundings as heat to guarantee an overall entropy increase.

In contrast, for an exothermic reaction having $\Delta S_{sys} > 0$, it is not necessary for any energy to be transferred to the surroundings to guarantee that $\Delta S_{univ} > 0$. For such a process, not only is the whole of ΔH free to do work; but the quantity $T\ \Delta S_{sys}$ is also available. Work in excess of the magnitude of ΔH is obtainable from such a process.

The "free energy" available from processes having varying signs and magnitudes of ΔH_{sys} and ΔS_{sys}.

TEMPERATURE DEPENDENCE OF DG

Experiment shows that neither ΔH nor ΔS for chemical reactions is particularly sensitive to temperature. Thus a reaction run at 400 K is expected to have nearly the same standard enthalpy and entropy changes as the standard values at 298 K. Both ΔH_R^o and ΔS_R^o can therefore be treated as constants within a fairly large range of temperature centered on 298 K. Since

$$\Delta G^o = \Delta H^o - T\Delta S^o,$$

/it follows that a plot of ΔG_R^o against T should be linear with

$$\text{Slope} = -\Delta S_R^o$$

$$\text{Intercept} = \Delta H_R^o$$

Example: Consider the simple physical process

$$H_2O(l) \rightarrow H_2O(g)\ 1\ \text{atm}$$

for which $\Delta H_R^o > 0$ and $DS_R^o > 0$. Predict the general form of a plot of ΔG_R^o versus T for the process, and discuss significant aspects of the plot.

Solution: Since the standard enthalpy and entropy for the process are both positive, the plot should have a positive intercept and a negative slope. We make the following observations from the plot:

- At low T, the TΔS term is small, so the ΔH term dominates; ΔG is positive;
- At high T, the TΔS term is large, and dominates; ΔG is negative;
- At the point at which the plot crosses the T axis, the

enthalpy and T-entropy terms are balanced. At the crossover T, $\Delta G = 0$, meaning that the enthalpy and entropy changes for vaporization of water to form water vapour at 1 bar are balanced at this temperature. This is of course the normal boiling point, 373 K.

Guideline: As T increases, all processes proceed spontaneously (become more favorable) in the direction that increases S.

We have previously stated that the atomization of methane is non-spontaneous at 298 K because the entropy increase is not sufficient to offset the large enthalpy increase that accompanies the rupture of the carbon-hydrogen covalent bonds.

$$CH_4(g) \rightarrow C(g) + 4H(g)\ \Delta H, DS > 0$$

Guideline 9 suggests, however, than an increase in temperature should favour atomization. Indeed, at very high temperature, such that $T\Delta S$ is more positive than ΔH, atomization becomes spontaneous. Increased kinetic energy (temperature) favors deaggregation.

Example: The Thermodynamics of a Rubber Band. We carry out a simple experiment with a rubber band, which produces a surprising result. A rubber band is stretched and allowed to come to air temperature. It is then allowed to spontaneously contract. Its temperature is measured immediately after the process and is found to be lower than the initial temperature; i.e., contraction of the rubber band is endothermic. Why is contraction endothermic? Why does the rubber band spontaneously contract?

Solution: We apply the First Law to the process. First, since the rubber band contracts very quickly, there is no time for heat flow from the surroundings into the rubber band.

Therefore,

$$q = 0$$

Second, since the rubber band contracts by itself with no assistance from outside, there is no work done.

Thus,

$$w = 0$$

The inevitable conclusion from the first law is that,

$$\Delta E = 0$$

The internal energy of the rubber band does not change when it contracts.

That the temperature of the rubber band decreases on contraction implies that the average kinetic energy of the rubber molecules decreases:

$$\Delta KE < 0$$

Since $\Delta E = \Delta KE + \Delta PE$, it follows that $\Delta PE > 0$. When the rubber band contracts, its potential energy increases in some way. Realise that this conclusion is forced on us by experiment (the observed decrease in T) and application of the First Law. It is possible to reach this conclusion without saying anything about the molecular nature of the rubber band. It is a conclusion arrived at by macroscopic observation. To go further—to understand the manner in which potential energy increases for the rubber band—requires that we switch to the molecular perspective.

Rubber is a polymer, which means that its molecules are like long strings. When the rubber band is stretched, so are the molecules. They align in the direction of stretching and thereby approach each other closely along their entire lengths. The intermolecular forces are strong, and PE is correspondingly low. When the rubber band contracts, the molecules relax and become tangled, like a mass of spaghetti, at right. In this state, molecules are in contact only where they cross over each other.

Intermolecular forces are weak, and PE is correspondingly high. According to this simple picture of the rubber molecules, intermolecular PE increases when the rubber band contracts. Since there is no time for heat to flow into the rubber band to supply the energy needed to overcome intermolecular forces, the rubber band gets the energy from itself; it converts molecular kinetic energy to potential energy, and cools.

Example: The spokes of a bicycle wheel are replaced by rubber bands, and the wheel is suspended above the ground as shown, supported by the axle. Initially the wheel is stationary (i.e., not rotating). A bank of heating lamps is placed behind the wheel, close to the rubber bands. When the lamps are turned on, which way will the wheel rotate?

Solution: The wheel will rotate counterclockwise. See if you can develop a reasonable explanation.

The qualitative and quantitative aspects of entropy and have applied the ideas to chemical and physical reactions. In the course of this development, entropy has emerged as an odd quantity. It is a state function for the system, like energy; yet it is very different from energy. When something happens in the world, whether it is a natural process like a volcanic eruption or a man-driven process like the manufacture of an automobile, energy may be changed in form, but it is not changed in amount: the total energy of the universe is constant (conserved). However, this is not true of entropy.

Entropy is created every time something happens! In fact, only events that create entropy are allowed. This gives a preferred direction — a one-wayness — to events that has led to the idea that entropy is in some way connected to the unidirectional flow of time. Further, the fact that entropy is continually being created means that it is not a conserved quantity for the universe. Yet we have defined it to be conserved for the system of interest. Therefore, it is not conserved for the surroundings. This is a very odd state function indeed. It is conserved for what we choose to define as the system, but it is not in general conserved outside of that system. The justification for dealing with entropy in this strange way is that it works in both the explanatory and the predictive senses. It is therefore very important to grapple and come to terms with the strangeness of entropy.

Supplement

We have seen that when $\Delta G^o < 0$ for a chemical or physical process, the process is spontaneous left to right from a starting point in which all reactants and products are in their standard states. When $\Delta G^o > 0$, the process is spontaneous right to left from a starting point in which all reactants and products are in their standard states. It is logical, then, that when $\Delta G^o = 0$, the process is equally spontaneous in both directions; it is in equilibrium under standard conditions. In this situation, reactants and products exist together indefinitely, each in its standard state, because the driving force for conversion of reactants to products is balanced by the driving force for conversion of products to reactants.

We learned that this situation is dynamic. At equilibrium, both conversions continue to take place at the molecular level. They take place at equal rates, however, so that from the macroscopic perspective, amounts of reactants and products do not change. The conversion of liquid water to water vapour is a process with which everyone has some familiarity, so we will use this process to explore the connection between equilibrium and $\Delta G^o = 0$.

The process is shown below:

$$H_2O(l) \Leftrightarrow H_2O(g)$$

We have used the double arrow to indicate that the process establishes dynamic equilibrium in a closed container. Suppose we are interested in determining the temperature at which liquid water is in equilibrium with water vapour at 1 atm pressure (i.e., the temperature at which the vapour pressure of water is 1 atm). We can estimate this temperature by taking advantage of the fact that $\Delta G^o = 0$ at this temperature. It can be shown that $\Delta H_R^o = 44.05$ kJ and $\Delta S_R^o = 118.7$ J/K for this process. At any particular temperature, it is true that,

$$\Delta G_R^o = \Delta H_R^o - T\Delta S_R^o.$$

Setting $\Delta G_R^o = 0$ and solving for T gives:

$$T = \Delta H_R^o/\Delta S_R^o$$

Thus the temperature at which liquid water exists in equilibrium with water vapour at 1 atm pressure is 370 K. At this temperature, the conversion of liquid to vapour and vapour to liquid are equally spontaneous.

We recognize that 370 K is the normal boiling point of water (actually, we do not obtain exactly 373 K because we have used values of the standard enthalpy and entropy for the process that are valid at 298 K. The values are somewhat different at 373 K). For a pure substance, the temperature at which DG° for conversion of liquid to vapour = 0 is the normal boiling point.

Chemical reactions carried out in closed systems also come to dynamic equilibrium, in which the forward reaction rate (conversion of reactants to products) and the reverse reaction rate (conversion of products to reactants) are equal. For a chemical reaction, $\Delta G_R{}^o = 0$ implies that reactants and products exist in equilibrium with each substance in its standard state. The temperature at which this situation exists can be estimated using the same approach used for phase equilibrium above. For example, reaction has $\Delta H_R{}^o = 57.2$ kJ and $\Delta S_R{}^o = 175.7$ J/K. Note that these values are consistent with bond breaking.

$$N_2O_4(g) \rightarrow 2NO_2(g)$$

$\Delta G_R{}^o = 0$ at a temperature such that,

$$T = \Delta H_R{}^o/DS_R{}^o, \text{ or } 326 \text{ K.}$$

At this temperature, we can have N_2O_4 and NO_2 in the same container, each with a partial pressure of 1 bar, with forward and reverse reaction tendencies exactly the same. In other words, the reaction is at equilibrium.

A very interesting aspect of equation deserves mention. Unless $\Delta H_R{}^o$ and $\Delta S_R{}^o$ have the same sign (both negative or both positive), equation yields a negative Kelvin temperature! The physical significance of this is that processes for which $\Delta H_R{}^o$ and $\Delta S_R{}^o$ have opposite signs can not have $\Delta G_R{}^o = 0$ at any temperature. They are either spontaneous under standard conditions at all temperatures (exothermic with positive entropy change) or non-spontaneous under standard conditions at all temperatures (endothermic with negative entropy change).

Our study of phase equilibrium revealed that two phases of a pure substance can be in equilibrium over a fairly wide range of temperature and pressure conditions. At 298 K, liquid water is in equilibrium with water vapour at 23.8 torr (0.0313 atm) in a closed container.

$$H_2O(l) \rightarrow H_2O(g)$$

$$P = 0.0313 \text{ atm}$$

This process is equally spontaneous in both directions; consequently it must be true that $\Delta G = 0$. Because the pressure of water vapour is not 1 atm, however, the water vapour is not in its standard state, and DG is not a standard value. It is written without the superscript o to indicate this. The standard free energy change for process can be calculated in the usual way:

$$\Delta G_R^o = \Delta H_R^o - T\Delta S_R^o = 8.7 \text{ kJ}.$$

This tells us that the process in which water vapour at 1 atm is formed from liquid water at 25 °C is nonspontaneous; steam at 1 atm pressure spontaneously condenses at 25 °C.

Chapter 4

Chemical Reactions

THERMOCHEMISTRY

Thermochemistry is the subject that deals with the heats involved in chemical reactions. A typical chemical reaction might have a form similar to the following hypothetical chemical reaction:

$$a\,\mathrm{A} + b\,\mathrm{B} \rightarrow c\,\mathrm{C} + d\,\mathrm{D}. \qquad (1)$$

(In Equation (1) the upper case letters stand for elements or compounds and the lower case letters stand for small whole numbers which balance the reaction. You would read this as saying, "*a* moles of A reacts with *b* moles of B to give *c* moles of C and *d* moles of D.")

A chemical reaction is a process just like any other thermodynamic process. It has an initial state (the reactants) and a final state (the products). We can calculate the changes in internal energy, enthalpy, and so on for the reaction. For example,

$\Delta U = U_{\text{products}}$ ″ $U_{\text{reactants}}$

and

$\Delta H = H_{\text{products}}$ ″ $H_{\text{reactants}}$

One thing is sometimes not made very clear. "Reactants" and "products" in these equations means that the reactants and products are separated, isolated, and pure. Furthermore, the reactants and products are all at the same temperature and pressure. So, for example, the ΔH above is the enthalpy of *c* moles of C (isolated and pure in its own container at temperature, *T*, and pressure, *p*) plus the enthalpy of *d* moles of D (isolated and pure in its own container at temperature, *T*, and pressure, p) minus the enthalpy of a moles of A (isolated and pure in its own container at temperature, T, and pressure, p) minus the enthalpy of b moles of B (isolated and pure in its own container at temperature, T, and pressure, p).

From time to time we will add a superscript ^{o} to H or U to indicate that reactants and products are in their "standard states." That is, they are in their most stable state at T and p. For example, the standard state of water at 25°C and 1 atm pressure is liquid water. If our reaction takes place at constant V, as in a bomb calorimeter, $dV = 0$ and,

$$\Delta U_V = q_V.$$

If the reaction takes place at constant p, as in open to atmospheric pressure, $dp = 0$ and,

$$\Delta H_p = q_p.$$

If $\Delta H_p < 0$ we say that the reaction is exothermic. That is, the system gave heat to the surroundings. On the other hand, if $\Delta H_p > 0$ we say that the reaction is endothermic. The system absorbed heat from the surroundings.

From the definition of enthalpy we find that

$$DH = DU + \Delta(pV), \qquad (2)$$

where,

$$\Delta (pV) = (pV)_{\text{products}} - (pV)_{\text{reactants}}.$$

For liquids and solids $\Delta(pV)$ is quite small. $\Delta(pV)$ is not necessarily small for gases, but we can get a reasonable estimate for this quantity by approximating the gases as ideal. Then,

$$\Delta (pV)_{\text{gas}} \approx (pV)_{\text{gas products}} - (pV)_{\text{gas reactants}} \approx n_{\text{gas products}} RT - n_{\text{gas reactants}} RT = RT\Delta n_{\text{gas}},$$

where Δn_{gas} is the difference in the number of moles of gaseous products and reactants. Using this approximation we get

$$DH = DU + RT\Delta n_{\text{gas}}. \qquad (3)$$

However, we have to be careful how we understand this equation because the conditions of the reaction must be the same on both sides of the equation. Since ΔH is the heat we measure if the reaction is run and constant pressure ($\Delta H_p = q_p$) and ΔU is the heat we measure if the reaction is run at constant volume ($\Delta U_V = q_V$), it is tempting (and common) to write Equation (3) as,

$$q_p = q_V + RT\Delta n_{\text{gas}}. \qquad (4)$$

However, Equation (4) cannot be rigorously true since the q's refer to different conditions, one at constant p and one at constant V. We can get an indication whether or not Equation (4) is a good approximation with the following example:

Consider the reaction:

$$2\ C(s) + O_2(g) \rightarrow 2\ CO(g), \qquad (5)$$

run at $dV = 0$.

$$\Delta(pV) \approx \Delta(pV)_{\text{gas}} \approx RT\Delta n_{\text{gas}} = RT.$$

The heat measured is q_V. Now let us find a way to measure q_p Consider the following two steps

$$2\ C(s) + O_2(g) \rightarrow 2\ CO(g) \rightarrow 2\ CO(g) \qquad (6)$$

$$p_1, V_1, T, q_V\ p_2, V_1, T, \Delta H_2\ p_1, V_2, T.$$

The first step is the constant volume reaction we had before with $\Delta U_V = q_V$. Notice that the pressure increases. The second step takes the product of the constant volume reaction and reduces the pressure back to the original pressure. We call the heat for this step ΔH_2.

So it is rigorously true that,

$$\Delta H_p = \Delta H_V + \Delta H_2 = q_V + RT + \Delta H_2. \qquad (7)$$

However, the ΔH_2 term is the enthalpy for the expansion of a gas at constant temperature. If the gas is ideal this term is zero. For real gases this term would be very small, so we make a negligible error is neglecting it. So to pretty good approximation we can use the equation,

$$q_p = q_V + RT\Delta n_{\text{gas}}. \qquad (4)$$

HESS' LAW

Hess' law states that if you add or subtract chemical reaction equations you can (must) add or subtract their corresponding ΔH's or ΔU's to get ΔH or ΔU for the overall reaction. For example, if we add the reaction

$$a\,A + b\,B \rightarrow c\,C + d\,D. \qquad \Delta H_1, \Delta U_1$$

to the reaction,

$$e\,E + f\,F \rightarrow g\,G + h\,H \qquad \Delta H_2, \Delta U_2$$

to get,

$$A + b\,B + e\,E + f\,F \rightarrow c\,C + d\,D + g\,G + h\,H,$$

Then, for the overall reaction,

$\Delta H = \Delta H_1 + \Delta H_2$ and $\Delta U = \Delta U_1 + \Delta U_2$.

The great utility of Hess' law is that we don't have to tabulate ΔH for every possible reaction. We can get ΔH for a particular reaction by adding and subtracting ΔH's for a much smaller set of reactions, called formation reactions. We define $\Delta_f H^\circ$ for a compound to be the enthalpy of the reaction: pure, isolated elements, in their standard (most stable) states →one mole of compound in its standard state.

For example, the heat of formation of liquid water is defined as ΔH° for the reaction,

$$H_2(g) + 1/2\ O_2(g) \rightarrow H_2O(l).$$

By definition, then, the heat of formation for an element in its most stable state is zero.

To obtain ΔH° for the hypothetical reaction, Equation (1), we add and subtract the appropriate heats of formations,

$$\Delta_r H^\circ = c\Delta_f H_C{}^\circ + d\Delta_f H_D{}^\circ - a\Delta_f H_A{}^\circ - b\Delta_f H_B{}^\circ \qquad (8)$$

For example, ΔH° for the reaction,

$$CH_4(g) + 2\ O_2(g) \rightarrow CO_2(g) + 2\ H_2O(l), \qquad (9)$$

is given by,

$$\Delta_r H^\circ = \Delta_f H^\circ{}_{CO_2(g)} + 2\Delta_f H^\circ{}_{H_2O(l)} - \Delta_f H^\circ{}_{CH_4(g)}$$

$$1\,\text{mol}\times(-393.51\frac{\text{kJ}}{\text{mol}}) + 2\,\text{mol}\times(-285.83\frac{\text{kJ}}{\text{mol}}) -$$

$$1\,\text{mol}\times(-74.81\frac{\text{kJ}}{\text{mol}})$$

$$= -890.36\text{kJ}.$$

We don't always write out explicitly that the coefficients which balance the reaction have units, but we have done so here to make it clear that these numbers have units. This will become an issue later.

Δ H AT OTHER TEMPERATURES

Tables of heats of formation usually give data for reactions at 25°C. We frequently need to know the heat of a reaction at a temperature other than 25°C. If we know the heat capacities at constant pressure we can calculate the heat of reaction at a temperature other than 25°C. We use the following chain of reasoning.

We know that,

$$\left(\frac{\partial H}{\partial T}\right)_p = C_p$$

then,

$$\left(\frac{\partial H^\circ}{\partial T}\right)_p = C_p^o$$

and,

$$\left(\frac{\partial \Delta H^\circ}{\partial T}\right)_p = \Delta C_p^o, \text{ (10a, b, c)}$$

where $\Delta C_p{}^o$ is defined for our hypothetical chemical reaction, Equation (1) as,

$$\Delta C_p^o = cC^\circ{}_{pc} + dC^\circ{}_{pD} - aC^\circ{}_{PA} - bC^\circ{}_{PB} \quad (11)$$

Prepare Equation 10c for integration as

$$d\Delta H^\circ = \Delta C_p^o\, dT, \quad (12)$$

and integrate,

$$\Delta H^\circ(T_2) = \Delta H^\circ(T_1) + \int_{T_1}^{T_2} \Delta C_p^o dT. \quad (13)$$

For very accurate work we will have to use the temperature dependent heat capacities in Equations (11) and (13), but more often than not we can regard the heat capacities as approximately constant over the temperature range, so that Equation (13) becomes,

$$\Delta H^\circ(T_2) = \Delta H^\circ(T_1) + \Delta C_p^o(T_2 - T_1). \quad (14)$$

As an example, let's calculate the heat of reaction for the reaction in Equation (9) at 95°C.

$\Delta C_p{}^o$ for the reaction in Equation (9) is given by,

$$\Delta_r C_p^o = C_{pCO_2(g)}^o + 2C_{pH_2O(l)}^o - C_{pCH_4(g)}^o - 2C_{pO_2(g)}^o$$

$$1mol \times (37.11 \frac{J}{Kmol}) + 2mol \times (75.291 \frac{J}{Kmol})$$

$$-1mol\times(35.35\frac{J}{Kmol})-2mol\times(29.355\frac{J}{Kmol})$$

$$=93.67\frac{J}{K}.$$

Then, Equation (14) gives,

$$\Delta H^{\circ}=(95^{\circ}C)=\Delta H^{\circ}(25^{\circ}C)+\Delta C_p^o(368.15-298.15)$$

$$=-890.36kJ+93.67\frac{J}{K}\times 70K\frac{1kj}{1000J}$$

= -883.8. kJ (16)

(Note: There is another way to do this problem. Calculate ΔH^{o} for cooling the reactants down to 25°C, calculate ΔH^{o} for the reaction at 25°C, calculate ΔH^{o} for heating the products back up to 95°C, and then add them up. Since H is a state function ΔH is independent of path. This method will also work if one of the components of the reaction has a phase change somewhere in the temperature range.

For example, if we let the new temperature be over 100°C we would have to account for the vapourization of the liquid water product. This is not hard to do, but requires some extra steps, including also the use of the heat capacity of water vapour. If the upper temperature in this problem is above 100°C we carry out the reaction in several steps:

Step 1 - Cool the reactants from the upper temperature to 25°C,

Step 2 - Run the reaction at 25°C,

Step 3 - Heat the product CO_2 from 25°C to the upper temperature,

Step 4 - Heat the liquid water from 25°C to 100°C,

Step 5 - Vaporize the water at 100°C,

Step 6 - Heat the water vapour from 100°C to the upper temperature.

The heat of reaction at the upper temperature is the sum of the ΔH's for all six steps. Again we have taken advantage of the fact that ΔH is independent of path.)

Δ H AS MAKING AND BREAKING BONDS

Breaking a chemical bond is an endothermic process. That is, you must put energy into the system to break the bond.

Forming a chemical bond is an exothermic process. The energy released in forming the bond goes into the surroundings. We can make an estimate of the ΔH for a chemical reaction by adding the bond energies of all the bonds broken and subtracting the bond energies of all the bonds formed.

$$\Delta H \approx BE_{\text{bonds broken}} \text{ " } BE_{\text{bonds formed}}.$$

Let's try it on the gas phase reaction,

$N_2 + 3\ H_2 \rightarrow 2\ NH_3$.

From the tables we find the following bond energies:

N–N 945 kJ

H–H 436 kJ

N–H 388 kJ.

In the reaction we break 1 N–N bond and 3 H–H bonds. We form 6 N–H bonds. The approximate ΔH is then,

$\Delta H\ H \approx 1 \times 945 + 3 \times 436 \text{ " } 6 \times 388 = -75$ kJ.

This can be compared to the actual value of " 92 kJ. The method is not super accurate, but it gives a ball-park answer and might be useful in cases where other data are not available. Further, however it demonstrates graphically that the heat of a reaction is related to the making and breaking of chemical bonds.

HEATS OF FORMATION OF IONS IN SOLUTION

When you look in a table of heats of formation you find values listed for ions in water solutions. That is, you will find an entry for species such as $Na^+(aq)$. It is fair to ask where these numbers come from. We know that in equilibrium chemistry it is impossible to prepare $Na^+(aq)$ ions in solution all by themselves. It is possible to prepare a solution that has both $Na^+(aq)$ and Cl^- (aq) ions, but not a solution that has only ions of one charge.

The heats of formation of solutions of soluble ionic compounds can be measured. That is, we can measure the heat of formation of HCl(*aq*). The heat of formation of HCl(*aq*) is defined as ΔH° for the reaction,

$1/2\ H_2(g) + 1/2\ Cl_2(g) \rightarrow HCl(aq)$. (17)

Since ionic compounds in solution are completely dissociated, it must be true that,

$$\Delta_f H^\circ_{HCl(aq)} = \Delta_f H^\circ_{H^+(aq)} + \Delta_f H^\circ_{Cl^-(aq)}. \quad (18)$$

We cannot know the heat of formation of either of the ions in solution, but we do know their sum.

By convention we arbitrarily set the heat of formation of the $H^+(aq)$ ion equal to zero. That is,

$$\Delta_f H^\circ_{H^+(aq)} = 0 \quad (19)$$

Then the heats of formation of all other aqueous ions can be determined relative to the heat of formation of the $H^+(aq)$ ion. As a start, we see that

$$\Delta_f H^\circ_{Cl^-(aq)} = \Delta_f H^\circ_{HCl(aq)} - \Delta_f H^\circ_{H^+(aq)}$$
$$\Delta_f H^\circ_{HCl(aq)} - 0$$
$$\Delta_f H^\circ_{HCl(aq)} \quad (20a, b, c)$$

This convention allows us to build up a table of heats of formation of aqueous ions. For example, from the measured value of the heat of formation of NaCl(aq) and the knowledge that,

$$\Delta_f H^\circ_{Nacl(aq)} = \Delta_f H^\circ_{Na^+(aq)} + \Delta_f H^\circ_{Cl^-(aq)} \quad (21)$$

we find that

$$\Delta_f H^\circ_{Na^+(aq)} = \Delta_f H^\circ_{NaCl(aq)} - \Delta_f H^\circ_{Cl^-(aq)}$$
$$= \Delta_f H^\circ_{Nacl(aq)} + \Delta_f H^\circ_{HCl(aq)} \quad (22a, b)$$

Continuing these procedures we can define the heats of formations of other aqueous ions,

$$\Delta_f H^\circ_{Br^-(aq)} = \Delta_f H^\circ_{HBr(aq)} + \Delta_f H^\circ_{H^+(aq)}$$
$$\Delta_f H^\circ_{HBr(aq)} - 0$$
$$\Delta_f H^\circ_{HBr(aq)^-} \quad (23a, b, c)$$

We now have enough data in our table to calculate the heat of formation of aqueous NaBr without measuring it.,

$$\Delta_f H^\circ_{NaBr(aq)} = \Delta_f H^\circ_{Na^+(aq)} + \Delta_f H^\circ_{Br^-(aq)}$$
$$= \Delta_f H^\circ_{Nacl(aq)} + \Delta_f H^\circ_{Hcl(aq)} + \Delta_f H^\circ_{HBr(aq)^-} \quad (24a, b)$$

If the heat of formation of $H^+(aq)$ were set to some value other than zero, it would have cancelled out of Equation (24b). In this manner an entire table of heats of formation of aqueous ions can be built with all values relative to the heat of formation of the $H^+(aq)$ ion.

EXACT AND INEXACT DIFFERENTIALS

We have mentioned, from time to time, that the quantities, *U*, *H*, and so on, are state functions, but that *q* and *w* are not state functions. This has various consequences. One consequence is that we can write things like ΔU and ΔH, but we never write *q* or *w* with a Δ in front them. A more important consequence is that in a process ΔU and ΔH are independent of path. That is, ΔU and ΔH depend only on the initial and final states. However, *q* and *w* do depend on the path one takes to get from the initial to the final state.

Another consequence is that the differentials, *dU* and *dH* are mathematically different, in some sense, from *dq* and *dw*. Some writers write *dq* and *dw* with a line through the *d* to indicate this difference. We have not chosen to use such a specialized notation, but expect that we all will be able to just remember that *dU* and *dH* are mathematically different, in some sense, than *dq* and *dw*.

We must now consider in detail the nature of this difference.

Let's think, for the moment, in terms of functions of the variables *x* and *y* and consider the differential,

$$df = 2xy^3dx + 3x^2y^2dy. \quad (1)$$

We ask the question, does there exist a function, $f = f(x,y)$ such that,

$$df = \left(\frac{\partial f}{\partial x}\right)_y dx + \left(\frac{\partial f}{\partial x}\right)_x dy. \quad (2)$$

In other words, does a function $f(x,y)$ exist such that,

$$\left(\frac{\partial f}{\partial x}\right)_y = 2xy^3$$

and

$$\left(\frac{\partial f}{\partial x}\right)_y = 3x^2y^2\,? \quad (3a, b)$$

Euler's test provides a way to see whether such a function,

$f(x,y)$ exists. Euler's test is based on the fact that for "nice" functions (and all of our functions are "nice") the mixed second derivatives must be equal.

That is,

$$\left(\frac{\partial^2 f}{\partial x \partial y}\right) = \left(\frac{\partial^2 f}{\partial y \partial x}\right). \tag{4}$$

(On the left-hand side we take the derivative with respect to y first and then take the derivative of the result with respect to x, and vice versa on the right-hand side.

Let's try Euler's test on our differential, df. If f exists then the Equations (3a) and (3b) are correct. Use Equations (3a, b) to obtain the proposed second derivatives,

$$\frac{\partial}{\partial y}\left(\frac{\partial f}{\partial x}\right)_y = \frac{\partial}{\partial y} 2xy^3 = 6xy^2$$

and

$$\frac{\partial}{\partial x}\left(\frac{\partial f}{\partial y}\right)_x = \frac{\partial}{\partial x} 3x^2y^2 = 6xy^2 \tag{5a, b}$$

The mixed second derivatives are equal. So we conclude that there exists a function, $f(x,y)$ (actually x^2y^3) such that Equations (1) and (2) are equal.

The differential df, in Equation (1), is called an "exact differential" for the very reason that a function, f, exists such that Equation (2) can be used to calculate it.

Now, let's consider the differential,

$$dg = 2x^2y^3dx + 3x^3y^2dy. \tag{6}$$

Is dg an exact differential? Use Euler's test to find out. If dg is exact then the coefficients of dx and dy are the respective partial derivatives of g. Euler's test would then compare

$$\frac{\partial}{\partial y} 2x^2y^3 = 6x^2y^2$$

$$\frac{\partial}{\partial y} 3x^3y^2 = 9xy^2. \tag{7}$$

These are not equal so that the putative second partial derivatives are not equal to each other. The differential, dg, is not exact and there does not exist a function, $g(x,y)$, such that dg gives Equation (6).

Both of the differentials, df and dg can be integrated from, say, x_1, y_1 to x_2, y_2. The integral,

$$\int df = f(x_2, y_2) - f(x_1, y_1). \quad (8)$$

depends only on the initial and final points because df is exact and the function f exists.

The differential dg can be integrated, but there is no equivalent to Equation (8) for the integral of dg because there is no function, $g(x,y)$ which gives Equation (6). The integral of dg would have to be carried out along some path and we would find that the value of the integral depends on the path as well as on the initial and final points.

So, what is the purpose of all this? We are getting ready to present the second law of thermodynamics. One of the consequences of the second law will be the demonstration that for a reversible process dq/T is exact. dq_{rev} is not exact, but dq_{rev}/T is exact. That means that dq_{rev}/T is the differential of some new function (a state function) whose integral is independent of path. We will call the new state function, S, and name it the "entropy."

The absolute temperature, T, is called an "integrating denominator" for dq_{rev}. That is, when we divide the inexact differential, dq_{rev}, by T the resulting differential becomes exact. Notice that the inexact differential, dg above, has an integrating denominator. The variable, x, is an integrating denominator for dg. You can see this by noticing that $dg/x = df$.

LAW OF THERMODYNAMICS

When we introduced the first law of thermodynamics we claimed that it is a statement of repeated observation elevated to the status of a law. No one has ever been able to make a machine that produces work out of nothing (a perpetual motion machine of the first kind), so we assume that no such machine can be made.

We then write this statement in mathematical language and begin deriving the consequences of the statement. Ultimately, the validation of the law comes from the experimental verification of the consequences.

The second law of thermodynamics is also a statement of

repeated observation (or perhaps better yet, a statement of some things that have never been observed).

Here are two things that have never been observed:

1. Heat has never been observed to move spontaneously from a cold body to a hot body.
2. Heat has never been observed to be converted entirely into work with no other result.

So the second law, in words, is just the statement that these two things are impossible. that is:

1. It is impossible for heat to move spontaneously from a cold body to a hot body with no other result.
2. It is impossible to convert heat quantitatively into work with no other result.

The latter statement is sometimes phrased: "It is impossible to make a perpetual motion machine of the second kind." (A perpetual motion machine of the second kind is a machine that converts heat into work without doing anything else. Imagine an ocean liner that scoops up liquid water out of the ocean, pulls the heat out of the water and uses it to power the ship, and dumps the left-over ice cubes out the back of the ship. Note that a perpetual motion machine of the second kind would not violate the first law. Energy would be conserved because any heat extracted would be converted into work.)

The second law is why automobiles have radiators. Someone might ask why we throw away all that energy that dissipates from the radiator.

Why not capture the energy and use it do decrease our gas mileage? The answer is that if you don't dissipate the heat the engine burns up, as you would quickly find out if you bypassed the radiator with a hose or if you drained the coolant from the radiator. In order to convert these word statements into mathematical statements we can use we will have to develop some apparatus. First we define the "heat engine." A heat engine is a cyclic process that absorbs heat from a heat bath and converts it into work. We shall see that in the cyclic process the engine also dissipates some heat to a heat bath at a lower temperature. A crucial feature of the heat engine is that it returns to its original state after each cycle.

That means that for each cycle of the engine itself, $\Delta H = 0$, $\Delta U = 0$, $\Delta T = 0$, and so on. Presumably, less heat is given back at the lower temperature than was absorbed at the upper temperature so that the difference can be used to supply work to the surroundings. If you run the engine backwards by providing an external power source you get a heat pump (or a refrigerator), that is, a machine that absorbs heat from a lower temperature heat bath and gives it back to a heat bath at a higher temperature. But it takes work to do this.

Our procedure will be as follows:

- Define and characterize a particular heat engine, the Carnot Cycle. The Carnot cycle is a heat engine operating between two heat baths, one at an upper temperature, which we shall call T_U and the other at a lower temperature, T_L. The Carnot cycle uses the expansion and compression of an ideal gas to convert heat into work.
- Define the efficiency, *e*, of a Carnot cycle.
- Assume that we can find a heat engine, operating between the same two temperatures, which has efficiency greater than a Carnot cycle efficiency and then show that this violates both of the word statements of the second law given above. This leads to the conclusion that no heat engine or cycle can have an efficiency greater than the efficiency of a Carnot cycle.
- The conclusion that no cycle can have an efficiency greater than a Carnot cycle will lead us to the further conclusion that the integral of dq_{rev}/T is independent of path. Therefore, the differential dq_{rev}/T must be exact, which means that it is the differential of some state function which we will call, *S*. That is, $dq_{rev}/T = dS$.
- Another cycle can have an efficiency *less than* the efficiency of a Carnot cycle. This will lead us to the conclusion that dq/T might be less than dS if there is some irreversibility in the process. Putting the two possibilities together we will conclude that, dq/T d"

dS. This is the mathematical statement of the second law of thermodynamics.

One further comment. We have seen processes where heat is converted into work before. The isothermal reversible expansion,

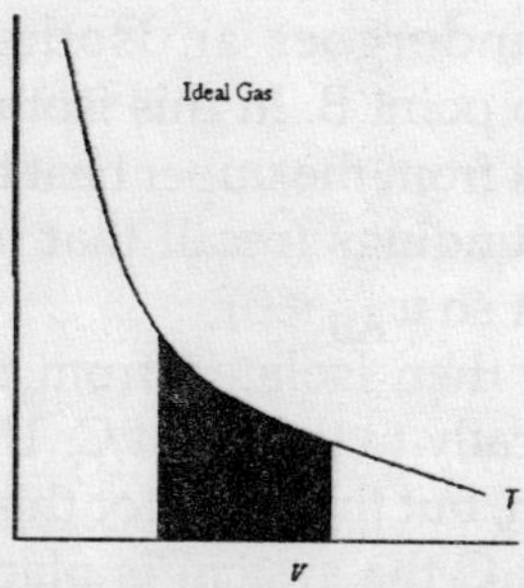

of an ideal gas is such a process. Recall that the work is the negative of the area under the curve. However, this is not a cyclic process. A change has taken place: the volume has increased and the pressure has decreased. In order to get more work out of the system you would have to expand the gas even further. This is *not* a process in which heat was converted into work and nothing else happened.

HEAT ENMGINES AND CARNOT CYCLE

A heat engine is a cyclic process that absorbs heat and does work on the surroundings. "Cyclic" means that the system returns to its initial state at the end of each cycle so that there is no permanent change in the system. The only cycle we will work with in this course is the Carnot Cycle which is shown below on a *p*-*V* diagram.

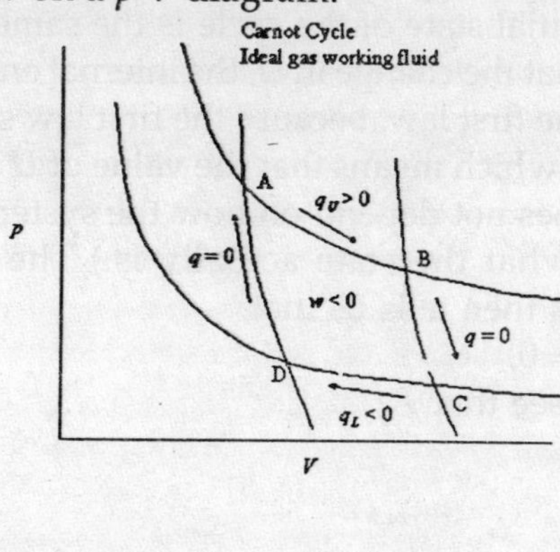

The curves labeled T_U and T_L are isotherms. T_U is the temperature of the upper temperature heat bath (reservoir) and T_L is the temperature of the lower temperature heat bath. The two steep curves, BC and DA, are adiabatic curves. The cycle begins at the point A.

The system undergoes an isothermal expansion at temperature, T_U, to point B. In this isothermal expansion the system absorbs heat from the upper heat bath ($q_U > 0$) and does work on the surroundings (recall that w is defined as work done on the system so $w_{AB} < 0$).

The system is then isolated from the heat bath and is expanded adiabatically to the point C. There is no heat in this adiabatic expansion, but the work for this step is also negative ($w_{BC} < 0$). At point C the system is placed in contact with a heat bath at T_L and undergoes an isothermal compression to point D.

For this segment of the cycle $q_L < 0$ and $w_{CD} > 0$ because the surroundings are now doing work *on* the system and heat is being dissipated to the heat bath at T_L. At point D the system is again isolated from the heat baths and compressed adiabatically to point A.

In this adiabatic compression the heat, of course, is zero and the work is positive ($w_{DA} > 0$)

We can analyse the Carnot cycle as follows: The heat, q, for the whole cycle is,

$$q = q_{AB} + q_{CD}, \quad (1)$$

and the work for the entire cycle is,

$$w = w_{AB} + w_{BC} + w_{CD} + w_{DA}. \quad (2)$$

Since the initial state of the cycle is the same as the final state we know that the change in U, the internal energy, is zero. (That's part of the first law, because the first law says that U is a state function, which means that the value of U for any state of the system does not depend on how the system got to that state, only on what the state actually is.) The first law of thermodynamics then tells us that,

$$\Delta U = q + w = 0, \quad (3)$$

from which we see that

$$q = -w \quad (4)$$

Notice that when the Carnot cycle is operating as a heat engine (going around clockwise), $w < 0$, so that $-w > 0$ and $q > 0$.

As we have said, the Carnot cycle is a heat engine. That means that we want it to absorb heat and convert that heat energy into work. We certainly should care about how much work we get out for the heat we absorb.

Thus we will define the efficiency, *e*, of a cycle as,

$$e = \frac{-w}{q_{AB}} \tag{5}$$

which, when combined with Equations 1 and 4 gives

$$e = \frac{q_{AB} + q_{CD}}{q_{AB}} \tag{6}$$

In order to conform to the usual notation we note that the A $\rightarrow$ B segment of the cycle is at the temperature of the upper heat bath, T_U, and the segment C $\rightarrow$ D is at the temperature of the lower heat bath, T_L.

Thus equation 6 becomes,

$$e = \frac{q_U + q_L}{q_U} = 1 + \frac{q_L}{q_U} \tag{7}$$

The A $\rightarrow$ B and C $\rightarrow$ D segments are an isothermal expansion of an ideal gas at T_U and an isothermal compression at T_L, respectively. For isothermal expansions and compressions of ideal gases.

Thus we can write,

$$q_U = -w_{AB} = +nRT_U In \frac{V_B}{V_A} \tag{8}$$

and

$$q_L = -w_{CD} = +nRT_L In \frac{V_D}{V_C} \tag{9}$$

Let's plug these heats into Equation 7, (right hand side) to get,

$$e = 1 + \frac{nRT_L In \frac{V_D}{V_C}}{nRT_U In \frac{V_B}{V_A}} \tag{10}$$

The nR cancels leaving,

$$e = 1 + \frac{T_L \ln \frac{V_D}{V_C}}{T_U \ln \frac{V_B}{V_A}} \tag{11}$$

If we are looking for an expression for efficiency in terms of temperatures Equation 11 won't do the job because of all the volumes in the equation.

However, there is another set of relations between the volumes from the fact that the paths B→ C and D→A are adiabatic and we know the volume relationships for adiabatic processes in an ideal gas. Recall that for an adiabatic expansion of an ideal gas we had the expression (using the temperatures, pressures, and volumes appropriate to our Carnot cycle here)

$$p_B V^{\gamma}{}_B = p_C V^{\gamma}{}_C \tag{12}$$

for the B→C segment of the cycle and a similar equation for the D→A segment. (Recall that γ is the "heat capacity ratio," C_p/C_V.) Let us eliminate the pressures from Equation 12 by inserting the value of pressure from the ideal gas equation of state, remembering that at points A and B the temperature is T_U and at points C and D the temperature is T_L.

$$\frac{nRT_U}{V_B} V^{\gamma}{}_B = \frac{nRT_L}{V_C} V^{\gamma}{}_C \tag{13}$$

Cancel nR from both sides and combine the volumes to get,

$$T_U V_B^{\gamma-1} = T_L V_C^{\gamma-1} \tag{14}$$

From which we obtain

$$\frac{T_U}{T_L} = \frac{V_C^{\gamma-1}}{V_B^{\gamma-1}} \tag{15}$$

Note that there is an equivalent expression for the D→A leg of the cycle,

$$\frac{T_U}{T_L} = \frac{V_D^{\gamma-1}}{V_A^{\gamma-1}} \tag{16}$$

From these two last equations we conclude that,

$$\frac{V_C^{\gamma-1}}{V_B^{\gamma-1}} = \frac{V_D^{\gamma-1}}{V_A^{\gamma-1}} \tag{17}$$

or

$$\frac{V_C}{V_B} = \frac{V_D}{V_A} \tag{18}$$

or, on rearranging,

$$\frac{V_A}{V_B} = \frac{V_D}{V_C} \tag{19}$$

We now apply this result to Equation 11 to get,

$$e = 1 + \frac{T_L In \frac{V_A}{V_B}}{T_U In \frac{V_B}{V_A}} \tag{20}$$

The two logarithms are the negatives of each other so we conclude that,

$$e = 1 - \frac{T_L}{T_U} = \frac{T_U - T_L}{T_U} \tag{21}$$

Notice that the highest efficiencies are obtained with a low temperature for the heat bath at T_L and that you could only obtain unit efficiency if the lower heat bath were at absolute zero. Since, in most cases, T_L is constrained by the surroundings engineers try to use the highest feasible operating temperature, T_U.

Real heat engines, of course, are not Carnot cycles, but the second law of thermodynamics requires that no heat engine operating between T_U and T_L can have an efficiency greater than a Carnot cycle efficiency. So the Carnot cycle provides an idealized upper limit to the efficiency of heat engines. Even though the Carnot cycle is idealized the general principles of heat engines remain the same.

If your lower temperature, T_L, is constrained (by design or operating considerations) you can increase the efficiency of your heat engine by increasing T_U. If your upper temperature, T_U, is constrained you can increase the efficiency by lowering T_L. You may have noticed that all of our temperatures are ideal gas temperatures. That is,they are measured on the ideal gas temperature scale.

It is possible, and desirable to use the Carnot cycle to define a "thermodynamic temperature scale." We will not do this here. Suffice it to say that by choosing a suitable reference temperature you can make the two scales identical.

SECOND LAW OF THERMODYNAMICS - TWO CYCLES

We have seen that the Carnot cycle, a reversible heat engine with an ideal gas working fluid, has an efficiency,

$$e = \frac{T_U - T_L}{T_U} \tag{1}$$

Is this the best we can do? Does there exist another cycle which has an efficiency greater than the efficiency of the Carnot cycle? We will now show that if such a cycle exists, with an efficiency greater than a Carnot cycle, then both of our word statements of the second law will be violated. That is, if a cycle (a heat engine working between the two heat reservoirs at T_U and T_L) exists with an efficiency greater than the efficiency of a Carnot cycle, then we can see heat spontaneously moving from a low temperature to a higher temperature with no other effect, *and* we will be able to convert heat quantitatively into work with no other effect. Let us call the cycle with higher efficiency the "better" cycle and just say about it that it has an efficiency, e', which is greater than e.

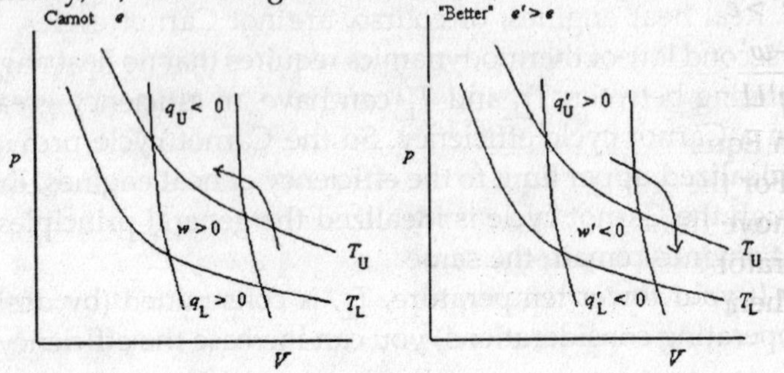

We will set up the two cycles such that the "better" cycle drives the Carnot cycle. That is, the "better" cycle will be operated as a heat engine and it will drive the Carnot cycle which then operates as a heat pump. We know that $\Delta U = 0$ for each cycle independently and for the sum of the two cycles.

Then,

$$\Delta U_{overall} = w + w' + q_U + q_L + q_U' + q_L' = 0, \qquad (2)$$

or

$$w + w' = -(q_U + q_L + q'_U + q'_L). \qquad (3)$$

We can adjust each cycle to alter the various q's and w's by changing where the adiabatic curves intersect the isotherms. Changing the positions of the adiabatic lines changes the area enclosed by the cycle and also changes the heats absorbed and released on the isothermal lines.

As we have said, in both of our two experiments the "better" cycle runs clockwise - as an engine, and the Carnot cycle runs counterclockwise as a heat pump. The engine drives the heat pump. Some or all of the work produced by the "better" cycle is used to run the (Carnot cycle) heat pump.

Experiment 1

Adjust the parameters of the two cycles such that $w' + w = 0$. That is, the work produced by the "better" cycle is entirely used up to drive the Carnot heat pump. This also means that there is no net work either on the system or on the surroundings.

Then,

$w = {}^{-}w'$,

which we can use in the expressions for the efficiencies,

$$e' > e$$

$$\frac{-w'}{q'U} > \frac{-w}{qU}. \qquad (4a, b)$$

In Equation 4b use the fact that $w =$ " w' on the left-hand-side. For the right-hand-side, use the fact that q_U is negative and move the negative side to the denominator so that both numerator and denominator are positive on the right.

Then,

$$e' > e$$

$$\frac{+w'}{q'_U} > \frac{-w'}{q'_U} > \frac{-w}{q_U} = \frac{w}{-q_U}$$

$$\frac{+w}{q'_U} > \frac{w}{-q_U}. \qquad \text{(4a, b, c)}$$

In Equation (4c) w is positive so we can divide both sides by w without affecting the inequality.

$$\frac{1}{q'_U} > \frac{1}{q_U}'$$

$$q'_U < -q_U, \qquad \text{(5,a b)}$$

since both q'_U and " q_U are positive. Then,

$$q'_U + q_U < 0. \qquad (6)$$

If you were to stand back and take a look at the overall effect of this experiment, including both cycles, you would see that, according to Equation (6), heat is being released into the heat bath at the upper temperature. This is heat that must have been absorbed from the heat bath at the lower temperature because of the first law and as indicated in Equation (3). However, we also know that there is no net work being done on the two systems.

The work of the "better" cycle is entirely used up driving the Carnot cycle. It appears, looking at the overall effect, that heat is spontaneously disappearing from the lower temperature heat bath and appearing in the upper temperature heat bath and there is no other effect. This violates our first word statement of the second law.

This time adjust the parameters of the experiment so that heat given to the lower temperature heat bath by the "better" cycle is exactly balanced by the heat absorbed at the lower temperature by the Carnot cycle.

That is, set

$$q_L + q'_L = 0 \qquad (7)$$

This time we find that,

$$e' > e$$

$$\frac{q'_U + q'_L}{q'_U} > \frac{q_U + q_L}{q_U}$$

$$1+\frac{q'_L}{q'_U}>1+\frac{q_L}{q_U}$$

$$\frac{q'_L}{q'_U}>\frac{q_L}{q_U}. \qquad \text{(8a, b, c, d)}$$

Invert (8d) to get

$$\frac{q'_U}{q'_L}<\frac{q_U}{q_L}$$

$$\frac{-q'_U}{-q'_L}<\frac{q_U}{q_L}. \qquad \text{(9a,b)}$$

Recall that, by construction of Experiment 2, we have from Equation (7),

$$-q'_L<q_L>0. \qquad (10)$$

Equation (9b) becomes,

$$\frac{-q'_U}{q_L}<\frac{q_U}{q_L},$$

$$-q'_U<q_U,$$

$$0<q'_U+q_U. \qquad \text{(11a, b, c)}$$

The net effect is that heat has been absorbed at the upper temperature and work has been done on the surroundings, but there was no net heat transferred to the lower temperature heat bath. This violates the second word statement of the second law.

If we regard the two cycles as one large heat engine, then that engine has no "radiator." It would be nice if such an arrangement would work, and from time to time people propose schemes which are designed to make it work, but it appears that nature does not allow it.

The initial assumption that we can find another cycle, with an efficiency, *e'*, better than the Carnot cycle efficiency, leads to a contradiction of both word statements of the second law. We are forced to conclude that no cycle can have an efficiency greater than the efficiency of a Carnot cycle.

That is,

$$e'\le e=\frac{T_U-T_L}{T_U} \qquad (12)$$

Equation (12) is useful and has a number of interesting and important consequences, but it does not yet provide us with the most significant result of the second law. On the next page we will use Equation (12) to define a new state function, S, called "entropy," and write the second law in a mathematical form which will give us enormous new calculating power.

THE EQUATION

We have seen that there are two word statements of the second law of thermodynamics. Both statements are just statements of universal observation. That is, no one has ever observed a violation of these statements and no one expects that a violation ever will be observed. As we did with the first law, we elevate these statements to the status of a "law" and assume that they are universally valid. Then we derive the consequences of this law, which can be checked out by experiment. So far, within the domain of its validity, no one has ever observed a violation of the second law, and its consequences are consistent with experimental observation.

Recall that the word statements of the second law are:

1. Heat does not move spontaneously from a cold body to a hot body with no other effect.
2. You can not convert heat quantitatively into work with no other effect.

The first statement is pretty obvious. Heat flows from hot bodies to cold bodies not the other way around. It would be very startling if all of the heat in your pencil spontaneously flowed into the eraser end so that the eraser melted and caught fire. The second statement is probably not so obvious unless you are an engineer, but it is one of the reasons that automobiles have radiators. In practice it means that you can't convert heat into work without dissipating some of the heat into a heat bath at a lower temperature than your heat source.

We demonstrated on the previous page that both of these statements lead to the conclusion that no heat engine can have an efficiency greater than a Carnot cycle. Recall that the efficiency of a cycle operating as a heat engine between two heat baths, one at an upper temperature, T_U, and the other at a lower temperature, T_L, can be written in several ways. For the Carnot cycle we can write,

$$e = \frac{-w}{q_U} = \frac{T_U - T_L}{T_U} = 1 - \frac{T_L}{T_U} = \frac{q_U + q_L}{q_U} = 1 + \frac{q_L}{q_U} \qquad (1)$$

Likewise, for some other cycle, which we will indicate by putting a prime on the heats and work - still operating between T_U and T_L - we can write,

$$e = \frac{-w'}{q'_U} = \frac{T_U - T_L}{T_U} = 1 - \frac{T_L}{T_U} = \frac{q'_U + q'_L}{q'_U} = 1 + \frac{q'_L}{q'_U} \qquad (2)$$

We are now considering the cycles to be operating as heat engines (as opposed to refrigerators, or heat pumps) - going around the cycle clockwise, so that w and w' are negative, q_U and q'_U are positive, and q_L and q'_L are negative. (We will need to know the signs of these quantities in order to carry out the algebra of inequality signs.)

Since we know from the second law that no cycle can have an efficiency greater than a Carnot cycle, we can write,

$$e' \le e \qquad (3)$$

(We might ask what circumstances might cause the efficiency of an engine to decrease. The most obvious answer is that irreversibility, for example, friction, would cause the efficiency to go down. If there is any irreversibility in the cycle its efficiency will be degraded.)

Using our various expressions for the efficiency, Equations 1 and 2, we can rewrite equation 3 as,

$$e' = 1 + \frac{q'_L}{q'_U} \le e = 1 - \frac{T_L}{T_U} \qquad (4)$$

or,

$$\frac{q'_L}{q'_U} \le \frac{T_L}{T_U} \qquad (5)$$

Multiply both sides of Equation 5 by " 1, which reverses the inequality,

$$\frac{-q'_L}{q'_U} \ge \frac{T_L}{T_U} \qquad (6)$$

In Equation 6 we have purposely associated the minus

sign with the q'_L which makes both the numerators and denominators of both sides positive. With all quantities in this equation positive we can cross-multiply at will without worrying about the direction of the inequality.

Cross-multiplying we get,

$$\frac{-q'_L}{T_L} \geq \frac{-q'_U}{T_U} \tag{7}$$

or

$$0 \geq \frac{q'_U}{T_U} + \frac{-q'_L}{T_L} \tag{8}$$

The equal sign holds when the cycle is fully reversible and "greater than" sign holds if there is any irreversibility in the cycle.

In the case where the cycle is fully reversibly, then,

$$\frac{q'_{U,rev}}{T_U} + \frac{-q'_{L,rev}}{T_L} = 0 \tag{9}$$

This latter equation makes it look like,

$$\sum_{cycle} \frac{q_{rev}}{T} = 0 \tag{10}$$

(Recall that two of the "legs" of the cycle were adiabatic so that there is no reversible heat on those legs.) We can generalize Equation 10 to an arbitrary closed path in p, V space by noting that we can fill an arbitrary closed path with a large number of small Carnot cycles (an infinite number in the limit of small cycles).

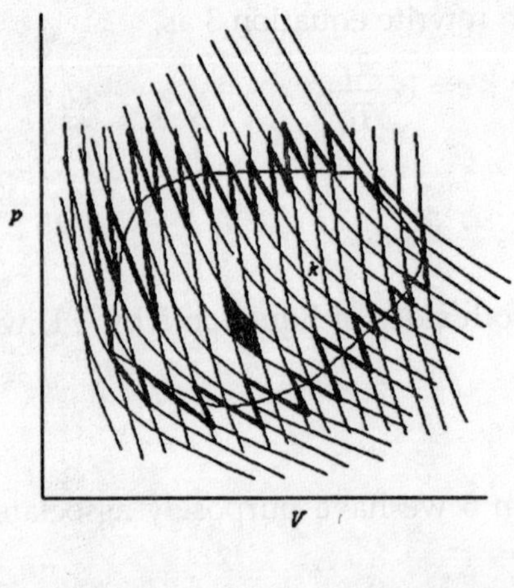

When we sum over all the cycles inside the heavy line the inside "lines" of the cycles (for example the blue filled Carnot cycle in the centre) will cancel because each inside line is traversed once clockwise and once counterclockwise. Therefore, only the outside lines (the heavy lines) remain. If we index the (small) cycles making up the complete cycle, by an index, *k*, where *k* ranges over all the cells inside the heavy line, this gives,

$$\sum_{all\,cycle} \frac{q_k}{T_k} = \sum_{outside\,lines,k} \frac{q_k}{T_k} \leq 0 \tag{11}$$

In the limit where we approximate our arbitrary closed path by making the isotherms and adiabatic lines closer and closer together to create an infinite number of cycles, the summation in Equation 11 becomes,

$$\oint_{closed\,path} \frac{dq}{T} \leq 0 \tag{12}$$

The equal sign holds when the entire path is reversibly,

$$\oint_{closed\,path} \frac{dq_{rev}}{T} \leq 0 \tag{13}$$

This equation says that the integral of reversible heat over temperature around a closed path is zero. Equation 13 is independent of the shape of the path - as long as it is a closed loop - which implies that,

$$\frac{dq_{rev}}{T} \tag{14}$$

is the differential of a state function. We will call this state function entropy and give it the symbol, *S*. Equation 14 becomes,

$$dS = \frac{dq_{rev}}{T}. \tag{15}$$

Combining Equations (12), (13), and (15) we get,

$$\oint_{closed\,path} \frac{dq}{T} \leq \oint_{closed\,path} \frac{dq_{rev}}{T} = \oint_{closed\,path} dS. \tag{16}$$

Since the closed path is arbitrary, Equation (16) must be

valid for any possible closed path. The only way this can be true is if the equation is true in its differential form.

That is,

$$\frac{dq}{T} \le \frac{dq_{rev}}{T} = dS, \tag{17}$$

which can be rewritten as simply

$$\frac{dq}{T} \le dS, \tag{18}$$

or, the best way to remember it,

$$dS \ge \frac{dq}{T}, \tag{19}$$

with the understanding that the equal sign holds when the process is reversible and the greater-than sign holds when the process is irreversible.

Equation (19) is the second law of thermodynamics in equation form. Equation (18) would work just as well, but most of us would prefer (19) because it is in the form of a definition of dS.

SECOND LAW APPLICATIONS

The second law of thermodynamics in equation form is,

$$dS \ge \frac{dq}{T}, \tag{1}$$

where the = sign holds when the process is reversible and the > holds when it is irreversible. Our first application of the second law will be to provide a thermodynamic definition of equilibrium.

We indicated very early in the course that at equilibrium none of the variables was changing in time. (Of course, a system in a steady state has to be excluded from this definition.)

Another possible definition of equilibrium was that all the variables have the values they would have at time equals infinity. Neither of these "definitions" provides an equation we can use to discuss equilibrium systems carefully. The second law provides us with a definition of equilibrium that we can use to derive equilibrium properties of thermodynamic systems. Consider a closed isolated system. In a closed isolated system $dq = 0$ and $dV = 0$ which implies that $dU = 0$. (Actually, we should really say that dw for all forms of work are zero, but for the time being we will only consider pV work.) Under the conditions of constant U and V the second law, Equation 1, becomes

$$dS_{UY} \geq 0. \qquad (2)$$

This may look simple, but it is a very profound statement. It says that in a closed isolated system (a system which us not being disturbed from the outside) any spontaneous change must increase the entropy.

In a closed, isolated system the entropy seeks a maximum. If you plot the entropy of a closed isolated system against some system variable any spontaneous change in the system must take the system to higher entropy.

If the system is not in equilibrium then,

$$dS_{UY} > 0,$$

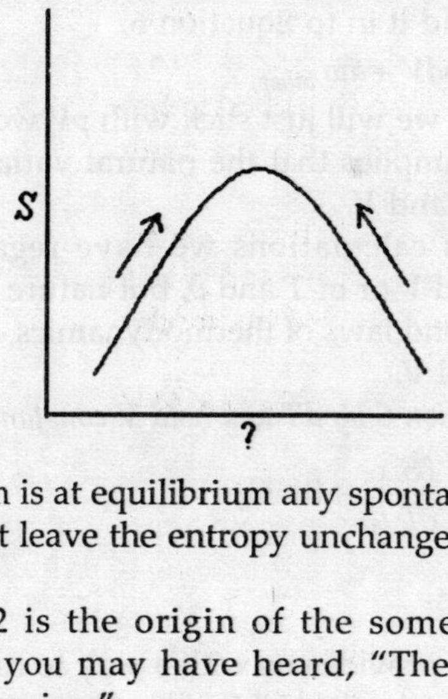

but if the system is at equilibrium any spontaneous change in the system must leave the entropy unchanged,

$$dS=0$$

Equation 2 is the origin of the somewhat arrogant statement that you may have heard, "The entropy of the universe is increasing."

If you regard the universe as a closed isolated system then the statement is probably true, even though it is difficult to take such broad statements concerning the nature of the universe seriously.

COMBINED FIRST AND SECOND LAWS

The first thing we must do is incorporate our new-found equation for the second law into what we already know. Going back to the first law, with pV work only, recal! that we can write,

$$dU = dq - pdV. \tag{3}$$

If we restrict our attention to reversible processes this becomes,

$$dU = dq_{rev} - pdV, \tag{4}$$

but we know from the second law that,

$$dq_{rev} = Tds. \tag{5}$$

Merging Equations 4 and 5 we arrive at what is called the combined first and second laws,

$$dU = TdS - pdV. \tag{6}$$

(Later on, when we want to include work other than pV work, we will add it in to Equation 6,

$$dU = TdS - pdV + dw_{other}, \tag{7}$$

but for now we will just stick with pV work.)

Equation 6 implies that the natural variables of internal energy, U, are S and V.

In previous calculations we have regarded U and a function of T and V or of T and p, but nature - in the form of the first and second laws of thermodynamics - gives us U as a function of S and V.

Divide Equation 6 by dT and hold V constant:

$$\left(\frac{\partial U}{\partial T}\right)_V = T\left(\frac{\partial S}{\partial T}\right)_V + 0 = C_V$$

$$\left(\frac{\partial S}{\partial T}\right)_V = \frac{C_V}{T}. \tag{8a, b}$$

Equation 8b provides us with a way to calculate entropy changes for a certain class of processes, namely processes at constant volume.

Set up Equation 8b for integration,

$$dS_V = \frac{C_V}{T}dT, \tag{9}$$

and calculate the entropy change for a constant volume process as,

$$\Delta S_V = \int_{T_1}^{T_2} \frac{C_V}{T} dT. \qquad (10)$$

Continuing to incorporate the second law into our set of thermodynamic tools, recall that

$$H = U + pV$$
$$dH = dU + pdV + Vdp$$
$$dH = TdS = pdV + pdV + Vpd$$
$$dH = TdS + Vdp. \qquad (11a, b, c, d)$$

(Note that equation 11d implies that the natural variables of H are S and p.) This time let's divide Equation 11d by dT and hold p constant,

$$\left(\frac{\partial H}{\partial T}\right)_p = T\left(\frac{\partial S}{\partial T}\right)_p + 0 = C_p$$
$$\left(\frac{\partial S}{\partial T}\right)_p = \frac{C_p}{T}. \qquad (12a, b)$$

We can use Equation 12b to calculate entropy changes for processes at constant pressure. Set up Equation 12b for integration,

$$dS_p = \frac{C_p}{T} dT, \qquad (13)$$

and integrate,

$$\Delta S_p = \int_{T_1}^{T_2} \frac{C_p}{T} dT. \qquad (14)$$

What about processes at constant T? We can calculate the entropy change for a process at constant temperature by rearranging Equation 5 for integration,

$$dS = \frac{dq_{rev}}{T}. \qquad (15)$$

The finite entropy change is,

$$\Delta S = \int \frac{dq_{rev}}{T}. \qquad (16)$$

If we now restrict consideration to processes at constant T Equation 16 becomes,

$$\Delta S = \int \frac{dqr_{,rev}}{T}$$

$$= \frac{1}{T} \int dqr_{,rev}$$

$$= \frac{qr_{,rev}}{T}. \tag{17}$$

Example

Calculate the entropy change in heating 1.00 mol of Al from 300 K to 500 K at constant pressure. The constant pressure heat capacity of Al is given to good approximation by,

$$C_p = 20.67 \frac{J}{Kmol} + 12.38 \times 10^{-3} T \frac{J}{K^2 mol}.$$

The calculation is

$$\Delta S_V = \int_{T_1}^{T_2} \frac{20.67}{T} dT + \int_{T_1}^{T_2} 12.38 \times 10^{-3} dT$$

$$= 20.67 \, In \frac{500K}{300K} + 12.38 \times 10^{-3} (500K - 200K)$$

$$= 13.04 \frac{J}{K}.$$

Example - An irreversible Process

Often we must calculate entropy changes for irreversibly processes. We don't know how to calculate entropy changes for irreversible processes, but it doesn't matter. Entropy is a state function so ΔS is independent of path. All we have to do is imagine a reversible path which will effect the same change and calculate the entropy change for the reversibly path.

Suppose we start with a 100. g block of Cu at 500 K and a 100. g block of Cu at 300 K. Bring the two blocks into thermal contact and heat will flow from the hotter block to the cooler block until they reach the same temperature.

This is definitely an irreversible process. However, let's imagine that we can reversibly cool the hot Cu block down to the final temperature and reversibly warm the cold Cu block up to the final temperature.

We can calculate the entropy change for both of these

processes from Equation (14). The total entropy change is just the sum of the two individual entropy changes. Note that we expect the entropy for this process to be positive because the process is spontaneous and the two Cu-block system can be regarded as isolated.

We need the following data:

$$C_{p,m,Cu} = 22.64\frac{J}{K}$$

$$FW_{Cu} = 63.5\frac{g}{mol}$$

First we must find the final temperature. We do this by recognizing that the heat lost by one Cu block is gained by the other Cu block. This is called "heat balance."

$$\frac{100g}{63.5\frac{g}{mol}}22.64\left(500 - T_{fin}\right) = \frac{100g}{63.5\frac{g}{mol}}22.64\left(T_{fin} - 300\right).$$

This is easily solved to obtain T_{fin} = 400 K. Then

$$\Delta S_1 = \frac{100g}{63.5\frac{g}{mol}}22.64\, In\frac{400K}{500K} = -7.95\frac{J}{K}$$

$$\Delta S_2 = \frac{100g}{63.5\frac{g}{mol}}22.64\, In\frac{400K}{500K} = +10.25\frac{J}{K}$$

$$\Delta S = \Delta S_1 + \Delta S_2 = 2.30\frac{J}{K}.$$

ENTROPY OF MIXING (IDEAL GASES)

Visualize that we have a container divided into two compartments. In one compartment we have n_1 moles of an ideal gas, gas 1, at pressure, p and temperature, T. In the other compartment we have n_2 moles of another ideal gas, gas 2, at the same p and T.

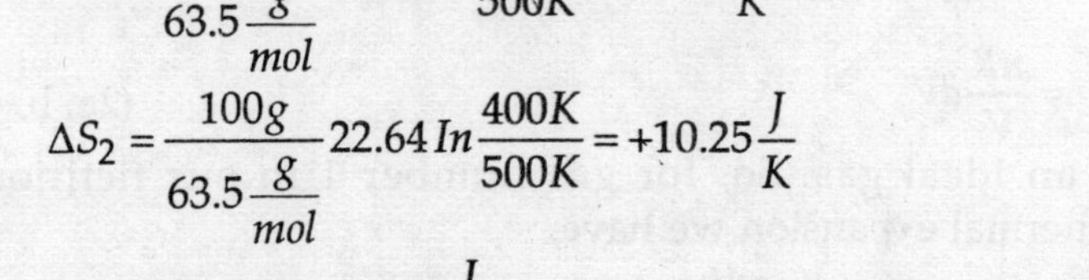

If we remove the partition the gases will begin to diffuse into each other and the system will eventually reach the state where both gases are uniformly distributed throughout the container. This is clearly an irreversibly process so that we would expect that the entropy would increase.

To calculate the entropy change we must find a reversible path to carry out the process, even if the path is fictitious. Imagine that we can devise a process that will expand one gas reversibly and isothermally, but leave the other gas undisturbed.

We know how to calculate the change in entropy for the reversible isothermal expansion of an ideal gas.

Recall that $dU = 0$ for the isothermal expansion of an ideal gas.

Then,

$$dU = TdS - pdV = 0 \tag{1}$$

So,

$$TdS = pdV$$

$$dS = \frac{p}{T}dV$$

$$= \frac{nR}{V}dV, \tag{2a, b, c}$$

for an ideal gas. So, for gas number 1 in our fictitious isothermal expansion we have,

$$\Delta S_1 = n_1 R\, In \frac{V_1 + V_2}{V_1}, \tag{3a}$$

and for gas number 2,

$$\Delta S_2 = n_2 R\, In \frac{V_1 + V_2}{V_2}. \tag{3b}$$

The total entropy change is the sum of these two individual entropy changes,

$$\Delta S_{mix} = \Delta S_1 + \Delta S_2 = n_1 R In \frac{V_1 + V_2}{V_1} + n_2\, R In \frac{V_1 + V_2}{V_2}. \tag{4}$$

Equation 4 could be used for calculations, but it is not in the form that we will see in other contexts. To obtain the usual form factor the R out of Equation 4 and invert the argument of the logarithms,

$$\Delta S_{mix} = R\left(n_1 In\frac{V_1}{V_1+V_2}+n_2\, In\frac{V_2}{V_1+V_2}\right). \tag{5}$$

Since we are doing a calculation for ideal gases, notice that the argument of the first logarithm in Equation 5 can be written,

$$\frac{V_1}{V_1+V_2}=\frac{\frac{n_1RT}{p}}{\frac{n_1RT}{p}+\frac{n_2RT}{p}}=\frac{n_1}{n_1+n_2}=X_1, \tag{6}$$

where X_1 is the mole fraction of component 1. There is an equivalent expression for the argument of the second logarithm. The entropy of mixing becomes,

$$\Delta S_{mix} = -R\left(n_1 In\, X_1 + n_2 In\, X_2\right). \tag{7}$$

Equation 7 is also suitable for calculations, but it is not yet in the standard form. To obtain the standard form write the total number of moles $n_1 + n_2$ as n and multiply and divide equation 7 by n. The result is

$$\Delta S_{mix} = -nR\left(X_1 In\, X_1 + X_2 In\, X_2\right). \tag{8}$$

Equation 8 is the same form that we will find when we derive an expression for the entropy of mixing ideal solutions and it is the same form that Shannon found for the "entropy of a message" in his famous series of papers on information theory.

If the two gases were not at the same initial pressure you would have to introduce some extra steps. Expand or compress one of the gases to bring it to the pressure of the other gas, mix the gases, and then compress or expand the mixture to bring it to the correct final volume and pressure.

If the two gases are not at the same temperature and pressure it is more complicated. You must find the final temperature using heat balance, reversibly cool and heat the two gases respectively to the same temperature, expand or contract on of the gases, mix them, and then expand or contract the mixture to the appropriate volume.

Equation 8 can easily be extended to more than two gases,

$$\Delta S_{mix} = -nR\left(X_1 In\, X_1 + X_2 In\, X_2 + X_3 In\, X_3 + ...\right). \tag{9}$$

Example: The molar entropy of dry air

The composition of dry air is approximately 78 per cent N_2, 21 per cent O_2, and 1 per cent Ar by volume (which is the same as mole per cent). What is the molar entropy of mixing of air?

$$\Delta S_{mix,air} = 1 \times mol \times 8.314 \frac{J}{K\,mol} (0.78\, In\, 0.78 + 0.21\, In\, 0.21 + 0.01\, In\, 0.01)$$

$$= 4.7 \frac{J}{K}.$$

ENTROPY MEASURES DISORDER

If we look at the processes we have seen which have positive entropy changes we can see that in each case an increase in entropy is associated with an increase in disorder.

An isothermal expansion gives the molecules more room to move around in, the molecules are less localized.

Increasing the temperature increases the average speeds of the molecules. The molecules are said to be more disordered in "velocity space" (or momentum space).

Mixing gases (or liquids) intersperses the molecules among each other increasing the disorder. Phase changes, such as going from a solid to a liquid or a gas, or from a liquid to a gas, increase the entropy because gases are more disordered than solids or liquids and liquids are more disordered than solids. For example, the entropy of fusion of 1.00 mol of ice at 273.15 (heat of fusion is 6.008 kJ/mol) is,

$$\Delta S_{fus} = \frac{6008 J}{273.15 K} = 22.0 \frac{J}{K}.$$

Vapourization of liquids has a large positive entropy of vapourization because gases are greatly disordered compared to liquids. A typical value is obtained from the vapourization of benzene at its boiling point. The heat of vapourization is 30.8 kJ at the boiling point, 353.1K. The entropy of vapourization is,

$$\Delta S_{Vap} = \frac{30800 J}{353.1 K} = 87.2 \frac{J}{K}.$$

It is interesting that the entropy of vapourization of many substances at their boiling points is close to about 86 J/K. (Water and helium are exceptions.) This phenomenon is called "Trouton's rule." It is easily understood on the basis of entropy being a measure of disorder. The vapourization process essentially "creates" a mole of disordered mclecules (the gas) from a mole of highly ordered molecules (the solid or liquid). The gases are all at one atmosphere pressure because we are at their normal boiling point.

Chapter 5

Adsorptions

PHASE

Consider a two phase system with a flat interface as shown in figure.

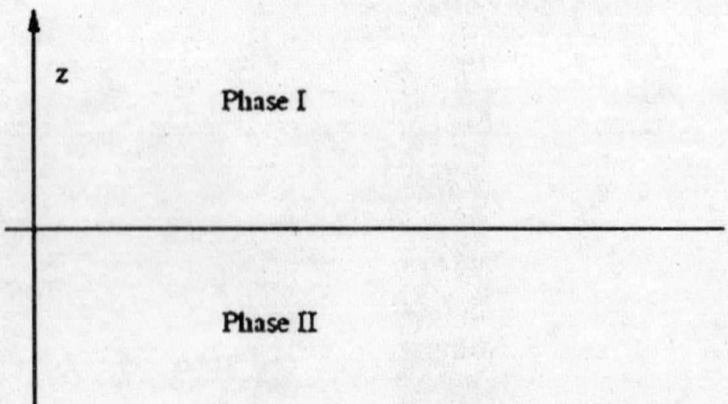

Since the two phases are in equilibrium in the presence of a gravity field, the pressure varies with height, z; namely,

$$p_1(z) = p_1(0) - \rho_1 g z \tag{1}$$

in phase 1, and

$$p_2(z) = p_2(0) - \rho_2 g z \tag{2}$$

in phase ρ1(2), where is the density in phase 1(2). Of course, at the interface (here assumed flat) the pressures must be equal and,

$$p_1(0) = p_2(0) = p_0. \tag{3}$$

Now put a capillary tube into the system as shown in figure.

The surface tension, γ, between phase 1 and phase 2 wants the pressures between the two phases to be unequal at the curved interface (why?). Specifically, for the convex surface shown in figure we know that

$$p_{2(-h)} - p_{1(-h)} = \frac{2\gamma}{R}. \tag{4}$$

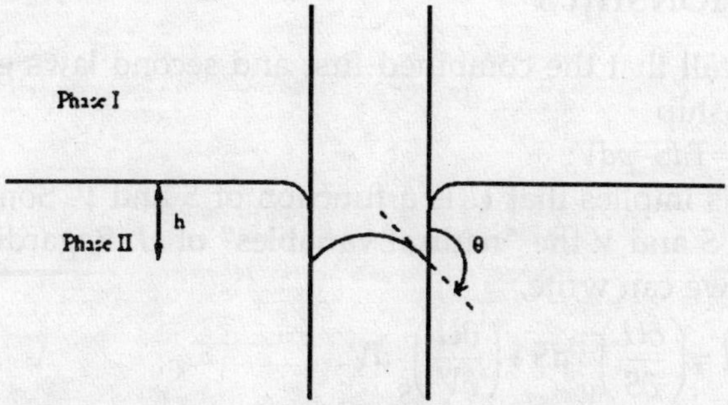

The only way in which the pressure difference can be achieved is by moving the curved interface to a new position in the fluid. (Changing the pressure in the gas or liquid from that of the surrounding material at the same height would cause material to flow from the high to low pressure regions until the pressures were equal at all points at the same height). Thus, by using our expressions for the pressure dependence in the phases at any given heights [Eqs. (1), (2), and (3)] in Eq. (4) we find that

$$p_0 + \rho_2 gh - (p_0 + \rho_1 gh) = \frac{2\gamma}{R} \tag{5}$$

or

$$h = \frac{2\lambda}{Rg(\rho_2 - \rho_1)} \approx \frac{2\gamma}{Rg\rho_2} \; for \; \rho_2 >> \rho_1, \tag{6}$$

where the approximate expression is useful for liquid-vapour interfaces. (If the interface is concave then the pressures in Eq. (4) must be switched and the sign of changes.)

Finally, if we write R, the radius of curvature of the interface, in terms of d, the radius of the tube and the contact angle, it follows that,

$$h = -\frac{2\gamma \cos(\theta)}{dg(\rho_2 - \rho_1)}$$

For a 45° contact angle in a capillary tube with a $1mm$ radius containing water $\gamma = 72.75\,dynes/cm$ (the capillary *rise* will be 1.05cm.

THERMODYNAMICS TOOLS

RELATIONSHIPS

Recall that the combined first and second laws give the relationship

$$dU = TdS - pdV. \tag{1}$$

This implies that U is a function of S and V. Sometimes we call S and V the "natural variables" of U. Regarding $U = U(S,V)$ we can write,

$$dU = \left(\frac{\partial U}{\partial S}\right)_V dS + \left(\frac{\partial U}{\partial V}\right)_S dV. \tag{2}$$

Comparing Equations 1 and 2 it is clear that,

$$\left(\frac{\partial U}{\partial S}\right)_V = T, \tag{3a}$$

and,

$$\left(\frac{\partial U}{\partial V}\right)_S = -p. \tag{3b}$$

These two equations can be regarded as thermodynamic definitions of T and p. Likewise, from the definition of enthalpy we wrote before that,

$$dH = TdS + Vdp. \tag{4}$$

Equation 4 implies that enthalpy is a natural function of S and p. Regarding $H = H(S,p)$ we can write,

$$dH = \left(\frac{\partial H}{\partial S}\right)_p dS + \left(\frac{\partial H}{\partial p}\right)_S dp \tag{5}$$

From Equations 4 and 5 it is clear that,

$$\left(\frac{\partial H}{\partial S}\right)_p = T, \tag{6a}$$

and,

$$\left(\frac{\partial H}{\partial p}\right)_S = V. \tag{6b}$$

Equations 6a and 6b give us another thermodynamic definition of T and a thermodynamic definition of V (which is curious since we have always regarded V as a purely mechanical variable).

HELMHOLTZ AND GIBBS FREE ENERGY

When we made the transformation from *U* to *H* by the Legendre transformation,

$$H=U+pV, \tag{7}$$

we remarked that *V* was not the most convenient independent variable. In the laboratory it is usually much easier to control pressure than it is to control *V*. Since both *U* and *H* are natural functions of entropy, it is fair to ask how convenient it is to have *S* as a variable. The answer is that it is not at all convenient to control entropy or to have entropy as an independent variable. We do not have a meter that reads entropy and we do not know how to hold entropy constant as we change some other variable. (Recall that we can control temperature, pressure, and volume.) So we make some more Legendre transformations.

We will not give an extensive discussion of Legendre transformations here, but we should point out that they are not arbitrary. You can't just pick any two variables you wish and put them together to make a Legendre transformation. We could make the Legendre transformation from *U to H* by adding the *pV* term to *U* only because *V* is related to p and *U* through Equation 3b. Using this as a guide it would seem reasonable to use Equation 3a to change the variable *S* to *T* in the function *U*, and use Equation 6a to change the variable *S* to *T* in the function *H*. Let's try it.

Define the Helmholtz free energy, *A*, as,

$$A \equiv U - TS. \tag{8}$$

Then,

$$\begin{aligned} dA &= dU - TdS - SdT \\ &= TdS - pdV - TdS - SdT \\ &= -SdT - pdV. \end{aligned} \tag{9a, b, c}$$

Equation 9c tells us that the Helmholtz free energy is a natural function of *T* and *V*. That is, $A = A(T,V)$. *T* and *V* are much more convenient variables than *S* and *V*. Regarding $A = A(T,V)$ we see that,

$$dA = \left(\frac{\partial A}{\partial T}\right)_V dT + \left(\frac{\partial A}{\partial V}\right)_r dV. \tag{10}$$

Now compare Equation 10 with Equation 9c to see that,

$$\left(\frac{\partial A}{\partial T}\right)_V = -S, \tag{11a}$$

and

$$\left(\frac{\partial A}{\partial V}\right)_r = -p. \tag{11b}$$

Equation 11a gives us a thermodynamic definition of entropy and 11b gives another thermodynamic definition of pressure. We now have a function of T and V, but we didn't much like V as an independent variable before so why should we like it any better now? Let's use the relationship 6a to define the Gibbs free energy, G, as,

$$G \equiv H - TS. \tag{12}$$

Actually, we can use any one of the three equivalent definitions,

$$G = H - TS$$
$$G = A + pV$$
$$G = U + pV - TS. \tag{13a, b, c}$$

Any one of Equations 13a, b, or c will give us the correct natural variables of G. Use Equation 12.

$$dG = dH - TdS - SdT$$
$$= TdS + Vdp - TdS - SdT$$
$$= -SdT + Vdp. \tag{14a, c, c}$$

From Equation 14c we see that the natural variables of G are temperature and pressure. Write

$$dG = \left(\frac{\partial G}{\partial T}\right)_P dT + \left(\frac{\partial G}{\partial p}\right)_r dp. \tag{15}$$

Comparing Equation 15 with Equation 14c we find that.

$$\left(\frac{\partial G}{\partial T}\right)_P = -S, \tag{16a}$$

and

$$\left(\frac{\partial G}{\partial p}\right)_r = V. \tag{16b}$$

Equation 16a gives us another thermodynamic definition of entropy and 16b another definition of volume.

Meaning of A and G

What do *A* and *G* mean and what are they good for? We said after the introduction of the first law (which introduced the internal energy, *U*) that we would be introducing three more functions that have units of energy. We now know that these functions are *H, A,* and *G*.

At the time we said that only *U* has a simple physical meaning - the sum of all the kinetic and potential energies of all the particles. There is no simple physical explanation for enthalpy and the two free energies. The best we can do is tell how they are used.

- Most simple-minded.

Set

$$dU = TdS - pdV + dw_{other} \quad (17)$$

Then, using the definition of Helmholtz free energy as we have done above we find that,

$$dA = -SdT - pdV + dw_{other} \quad (18)$$

For any process at constant temperature we have,

$$dA_T = -pdV + dw_{other} = dw_{all,reversible}. \quad (19)$$

That is, for a constant temperature process the Helmholtz free energy gives all the reversible work. For this reason the Helmholtz free energy is sometimes called the "work function." When a physicist says "free energy" without indicating Helmholtz or Gibbs, he usually means Helmholtz free energy.

Similarly, we can write,

$$dG = -SdT + Vdp \_ dw_{other}. \quad (20)$$

For a process at constant temperature *and* pressure we get,

$$dG_{T,p} = dw_{other} = dw_{all,non-pV}. \quad (21)$$

That is, for a process at constant temperature and pressure the change in Gibbs free energy gives all the reversible work except the *pV* work. This work might include electrical work, work creating surface area, and so on. Chemists do most of their reactions on the bench top at constant pressure. When a chemist says "free energy" she almost always means Gibbs free energy unless she specifically states otherwise.

- *More useful*: Two new criteria for equilibrium

Recall that the second law of thermodynamics,

$$dS \geq \frac{dq}{T}, \tag{22}$$

gives us the fundamental criterion for equilibrium. That is, in a closed isolated system entropy seeks a maximum,

$$dS_{U,V} \geq 0 \tag{23}$$

Although this is the fundamental definition of equilibrium it is not the most useful definition because we do not often work with closed isolated systems. More often we work with systems at constant temperature and either constant volume or constant pressure.

We can use the second law, Equation 22, and our new functions A and G to find criteria for equilibrium under these conditions.

Rewrite the second law, Equation 22 as follows:

$$TdS \geq dq$$

$$0 \geq dq - TdS, \tag{24}$$

or

$$dq - TdS \leq 0. \tag{25}$$

Now, going back to the original form of the first law with only pV work,

$$dU = dq - pdV, \tag{26}$$

and making the transformation to Helmholtz free energy we get,

$$\begin{aligned} dA &= dU - TdS - SdT \\ &= dq - pdV - TdS - SdT \\ &= (dq - TdS) - pdV - SdT. \end{aligned} \tag{27a, b, c}$$

For a process at constant temperature and volume we have,

$$dA_{TV} = dq - TdS \leq 0 \tag{28}$$

We conclude that for a process at constant temperature and volume the Helmholtz free energy seeks a minimum. Any spontaneous process in a system at constant T and V must decrease the Helmholtz free energy (if the system is away from equilibrium) or leave the Helmholtz free energy unchanged (if the system is at equilibrium).

By the same token, we can use the Gibbs free energy to discuss processes at constant temperature and pressure,

$$dG = dH - TdS - SdT$$
$$= dq + Vdp - TdS - SdT$$
$$= (dq - TdS) + Vdp - SdT, \qquad (29a, b, c)$$

$$dG_{T,p} = dq - TdS \leq 0. \qquad (30)$$

That is, at constant T and p the Gibbs free energy seeks a minimum. Any spontaneous process in a system at constant T and p must decrease the Gibbs free energy (if the system is away from equilibrium) or leave the Gibbs free energy unchanged (if the system is at equilibrium).

MAXWELL'S EQUATIONS

We now have the tools to derive some very useful relationships between thermodynamics variables. Maxwell's equations are based on the same principle as was Euler's test for exact differentials, namely that mixed second derivatives of "nice" functions must be equal. Applying this principle to our two new free energy functions we find,

$$\frac{\partial^2 A}{\partial V \partial T} = \frac{\partial^2 A}{\partial T \partial V}, \qquad (31)$$

but we already know the first derivatives of A from Equations 11a and 11b. So,

$$\frac{\partial^2 A}{\partial V \partial T} = \frac{\partial}{\partial V}\left(\frac{\partial A}{\partial T}\right)_V = \frac{\partial^2 A}{\partial T \partial V} = \frac{\partial}{\partial T}\left(\frac{\partial A}{\partial V}\right)_T,$$
$$\frac{\partial}{\partial V}\left(\frac{\partial A}{\partial T}\right)_V = \frac{\partial}{\partial V}(-S) = \frac{\partial}{\partial T}\left(\frac{\partial A}{\partial V}\right)_T = \frac{\partial}{\partial T}(-p),$$
$$\left(\frac{\partial S}{\partial V}\right)_T = \left(\frac{\partial p}{\partial T}\right)_V. \qquad (32a, b, c)$$

We obtain another similar equation from the Gibbs free energy,

$$\frac{\partial^2 G}{\partial p \partial T} = \frac{\partial^2 G}{\partial T \partial p}, \qquad (33)$$

which becomes, using Equations 16a and 16b,

$$\frac{\partial^2 G}{\partial p \partial T} = \frac{\partial}{\partial p}\left(\frac{\partial G}{\partial T}\right)_p = \frac{\partial^2 G}{\partial T \partial p} = \frac{\partial}{\partial T}\left(\frac{\partial G}{\partial p}\right)_T,$$

$$\frac{\partial}{\partial p}\left(\frac{\partial G}{\partial T}\right)_p = \frac{\partial}{\partial p}(-S) = \frac{\partial}{\partial p}\left(\frac{\partial G}{\partial p}\right)_T = \frac{\partial}{\partial T}(+V),$$

$$-\left(\frac{\partial G}{\partial p}\right)_T = \left(\frac{\partial V}{\partial T}\right)_p. \qquad \text{(34a, b, c)}$$

There are two more Maxwell's equations from dU and dH, but these are not as useful as the ones just derived. We will leave it to the reader to find the Maxwell's equations from dU and dH.

FIRST APPLICATION OF A MAXWELL'S EQUATION

As our first application of a Maxwell's equation we will derive the so-called thermodynamic equation of state which we stated without proof earlier. Write the combined first and second laws,

$$dU = TdS - pdV. \qquad (1)$$

Divide Equation 1 by dV and hold T constant to get,

$$-\left(\frac{\partial U}{\partial V}\right)_T = T\left(\frac{\partial S}{\partial V}\right) - p. \qquad (35)$$

Using the Maxwell's Equation, Equation 32c, to substitute for the entropy derivative we obtain,

$$\left(\frac{\partial U}{\partial V}\right)_T = T\left(\frac{\partial p}{\partial T}\right)_V - p. \qquad (36)$$

Equation 36 is the equation that was written down without proof at the time we were discussing the Joule expansion.

The other version of thermodynamic equation of state, based on H instead of U, will be left as an exercise for the reader.

We now have four interesting and useful derivatives of entropy,

$$\left(\frac{\partial S}{\partial V}\right)_T = \frac{C_V}{T}\left(\frac{\partial S}{\partial T}\right)_p = \frac{C_p}{T}$$

$$\left(\frac{\partial S}{\partial V}\right)_T = \left(\frac{\partial p}{\partial T}\right)_V \left(\frac{\partial S}{\partial p}\right)_T = -\left(\frac{\partial V}{\partial T}\right)_p .$$

There are two other derivatives of entropy which might prove useful,

$$\left(\frac{\partial S}{\partial p}\right)_V = \left(\frac{\partial S}{\partial T}\right)_V \left(\frac{\partial T}{\partial p}\right)_V$$

$$= \left(\frac{C_V}{T}\right)\left(\frac{\partial T}{\partial p}\right)_V . \qquad (37)$$

The other derivative,

$$\left(\frac{\partial S}{\partial V}\right)_p ,$$

will be left to the reader.

ADIABATIC COMPRESSIBILITY

The speed of sound in a gas depends on the "springiness" of the gas. That is, it depends on how the volume of the gas responds to changes in pressure. We have already seen one measure of this response, called the isothermal compressibility,

$$K_T = -\frac{1}{V}\left(\frac{\partial V}{\partial p}\right)_T . \qquad (1)$$

Equation one gives a parameter that determines how the gas responds to changes in pressure if the temperature remains constant.

Sir Isaac Newton assumed that the speed of sound was an isothermal process and used the parameter defined by Equation 1 to calculate the speed of sound in a gas. His answer did not agree with experiment.

It turns out that sound transmission in a gas is an adiabatic process rather than an isothermal process. The sound wave causes oscillations in pressure but the oscillations are fast enough that heat can not move from compressed regions to rarified regions in order to keep the temperature constant. Before the heat can be conducted away from the compressed regions the compression has moved on so that sound propagation is adiabatic.

We define the adiabatic compressibility as,

$$K_S = -\frac{1}{V}\left(\frac{\partial V}{\partial p}\right)_S . \tag{2}$$

We can calculate the adiabatic compressibility in terms of quantities that we already know (using the Euler cyclic rule twice, once the normal way and once in reverse).

$$k_S = -\frac{1}{V}\left(\frac{\partial V}{\partial p}\right)_S$$

$$= -\frac{1}{V}\left(-\frac{\left(\frac{\partial S}{\partial p}\right)_V}{\left(\frac{\partial S}{\partial V}\right)_p}\right)$$

$$= -\frac{1}{V}\left(-\frac{\left(\frac{\partial S}{\partial T}\right)_V\left(\frac{\partial T}{\partial p}\right)_V}{\left(\frac{\partial S}{\partial T}\right)_p\left(\frac{\partial T}{\partial V}\right)_V}\right), \tag{3a, b, c}$$

where we have used Euler's cyclic relation to go from Equation 3a to 3b and the chain rule in the denominator and numerator of 3b to go to 3c.

We already know that,

$$\left(\frac{\partial S}{\partial T}\right)_V = \left(\frac{C_V}{T}\right) \tag{4a}$$

and

$$\left(\frac{\partial S}{\partial T}\right)_p = \left(\frac{C_p}{T}\right), \tag{4b}$$

so

$$k_S = -\frac{1}{V}\left(-\frac{\left(\frac{C_V}{T}\right)\left(\frac{\partial T}{\partial p}\right)_V}{\left(\frac{C_p}{T}\right)\left(\frac{\partial T}{\partial V}\right)_p}\right)$$

$$= \frac{C_V}{C_p}\frac{1}{V}\left(\frac{\left(\frac{\partial T}{\partial p}\right)_V}{\left(\frac{\partial T}{\partial V}\right)_p}\right). \qquad \text{(5a, b)}$$

In Equation 5b we can use Euler's cyclic rule in reverse to write

$$k_S = \frac{C_V}{C_p}\frac{1}{V}\left(\frac{\partial V}{\partial p}\right)_T$$

$$k_S = \frac{C_V}{C_p}k_S. \qquad \text{(6a,b)}$$

Since $C_p > C_V$ The isothermal compressibility is always greater than the adiabatic compressibility.

For a monatomic ideal gas, where $C_p = 5nR/2$ and $C_V = 3nR/2$ we see that,

$$kS = \frac{3}{5}k_T. \qquad (7)$$

ADIABATIC GAS EXPANSION REVISITED

Early in the course we derived the equation for the adiabatic expansion of an ideal gas,

$$pV^\gamma = cons\tan t,$$
$$p_1V_1{}^\gamma = p_2V_2{}^\gamma, \qquad \text{(8a, b)}$$

where,

$$\gamma = \frac{C_p}{C_V}. \qquad (9)$$

You may recall that the derivation was sort of "roundabout." Here we would like to use some of our new thermodynamics tools to provide a much more direct derivation.

The goal is to discover how the pressure changes with volume if entropy is held constant. That is, we would like to find the derivative,

$$\left(\frac{\partial p}{\partial S}\right)_S.$$

Then we can integrate this derivative to find an expression for p as a function of V. First, let's find the derivative.

Notice that this derivative is just the reciprocal of the derivative in Equation 2 so most of the work has already been done. Using the same procedures we used above we fin that,

$$\left(\frac{\partial p}{\partial V}\right)_S = -\frac{\left(\frac{\partial S}{\partial V}\right)_p}{\left(\frac{\partial S}{\partial p}\right)_V}$$

$$= -\frac{\left(\frac{\partial S}{\partial T}\right)_p \left(\frac{\partial T}{\partial V}\right)_p}{\left(\frac{\partial S}{\partial T}\right)_V \left(\frac{\partial T}{\partial p}\right)_V}$$

$$= -\frac{\frac{C_p}{T}\left(\frac{\partial T}{\partial V}\right)_p}{\frac{C_V}{T}\left(\frac{\partial T}{\partial p}\right)_V} = \frac{C_p}{C_V}\left(\frac{\left(\frac{\partial T}{\partial V}\right)_p}{\left(\frac{\partial T}{\partial p}\right)_V}\right)$$

$$\left(\frac{\partial p}{\partial V}\right)_S = \gamma\left(\frac{\partial p}{\partial V}\right)_T. \qquad \text{(10a, b, c, d, e, f)}$$

Equation 10f is a general thermodynamic relationship. It contains no approximations. To proceed further we must decide what material we want to consider. The equation we were trying to derive was based on the ideal gas for which,

$$\left(\frac{\partial p}{\partial V}\right)_T = -\frac{nRT}{V^2}$$

$$= -\frac{p}{V}. \qquad \text{(11a, b)}$$

So,

$$\left(\frac{\partial p}{\partial V}\right)_S = -\gamma\frac{p}{V}. \qquad \text{(12)}$$

Set up Equation 12 for integration, (don't forget to put all the p stuff on one side and all the V stuff on the other)

$$\frac{dp}{p} = -\gamma \frac{dV}{V}, \tag{13}$$

and integrate between p_1, V_1 and p_2, V_2,

$$\int_{p_1}^{p_2} \frac{dp}{p} = -\gamma \int_{V_1}^{V_2} \frac{dV}{V}, \tag{14}$$

where we have made the approximations that γ is independent of volume and factored it out of the integral. Equation 14 integrates to

$$In \frac{p_2}{p_1} = -\gamma In \frac{V_2}{V_1}$$

$$= +\gamma In \frac{V_1}{V_2}$$

$$In \frac{V_1^{\gamma}}{V_2^{\gamma}}, \tag{15a, b, c}$$

from which we conclude that

$$\frac{p_2}{p_1} = \frac{V_1^{\gamma}}{V_2^{\gamma}}, \tag{16}$$

or

$$p_1 V_1^{\gamma} = p_2 V_2^{\gamma}. \tag{16b}$$

Just for the record, we add that the correct formula for the speed of sound is

$$v_{sound} = \frac{1}{\sqrt{\rho k_S}}, \tag{17}$$

where ñ is the density of the gas (in kg/m^3).

GIBBS FREE ENERGY

We have seen that the Gibbs free energy, G, seeks a minimum at constant T and p. The question now is does this have anything to do with chemical reactions. We define the change in Gibbs free energy for a chemical reaction as,

$$\Delta_r G^{\circ} = G_{products} - G_{reac\tan ts}. \tag{1}$$

It is understood in Equation 1 that both the reactants and the products are pure, isolated, in their standard states, and at the same temperature and pressure.

When we first introduced heats of reaction we said that it is not feasible to tabulate the ΔH° for every possible reaction. Instead, we tabulate heats of formation for compounds and use Hess' law to find the heat of a reaction involving those compounds.

In the same manner we define the Gibbs free energy of formation for a compound as ΔG° for the reaction:

Pure isolated elements in their standard states $\rightarrow$ one mole of pure compound in its standard state.

For example, $\Delta_f G^{\circ}$ for $CO_2(g)$ at 25° and 1 atm is defined to be the ΔG for the reaction:

$C(s, graphite) + 2\, O_2(g) \rightarrow CO_2(g)$.

Then, using Hess' law, the $\Delta_r G^{\circ}$ for some arbitrary reaction, say

$$a\,A + b\,B \rightarrow c\,C + d\,D, \tag{2}$$

is

$$\Delta_r G^{\circ} = c\,\Delta_f G_C{}^{\circ} + d\,\Delta_f G_D{}^{\circ} - a\,\Delta_f G_A{}^{\circ} - b\,\Delta_f G_B{}^{\circ}. \tag{3}$$

Rarely one might want to run a reaction at constant volume and, therefore, would need the change in the Helmholtz free energy. We can obtain $\Delta_r A^{\circ}$ using the same approximations we used to obtain ΔU from ΔH. That is, from

$$G = A + pV$$

we see that,

$$\Delta A = \Delta G - \Delta(pV).$$

Using, as before, the approximation that the change in the pV product for liquids and solids is small and the approximation that the gases are ideal, we obtain the equation,

$$\Delta_r A^{\circ} \approx \Delta_r G^{\circ} - RT\Delta n_{gas}. \tag{4}$$

PROCESSES AT CONSTANT TEMPERATURE

Knowing that $G = H - TS$, and that reactions are constant temperature process, we can relate the Gibbs free energy of a reaction to the enthalpy of a reaction. That is,

$$G = H + TS,$$

$$\Delta G = \Delta H + \Delta(TS)$$

$$\Delta G_T = \Delta H_T - T\Delta S_T. \tag{5a, b, c}$$

We have written a subscript, *T*, in Equation 5c to indicate that the equation is only valid at constant *T*, but it is unusual to include the *T*. Usually we see,

$$\Delta G = \Delta H - T\Delta S, \qquad (6)$$

and it is left to the reader to understand that the equation is only valid at constant *T*.

Recall that at constant *T* and *p* the Gibbs free energy seeks a minimum. That means that we can use $\Delta_r G^o$ to tell whether or not a reaction will proceed spontaneously as written:

If $\Delta_r G^o > 0$ then the reaction will not go as written (the reverse reaction will go) if $\Delta_r G^o < 0$ then the reaction will go as written.

Sometimes people say that $\Delta_r G^o$ is a measure of the "driving force" of a chemical reaction. Although the use of the word "force" is probably not appropriate, the statement conveys the correct idea that $\Delta_r G^o$ tells us whether or not a given reaction will really run spontaneously. (You can probably write and balance a large number of reactions, but just because you can write and balance a reaction is no guarantee that it will run.)

We see from Equation 6 that the "driving force" of a chemical reaction has two components:

ΔH is the drive toward stability. When $\Delta H < 0$ the products are more stable than the reactants (and vice versa).

ΔS is the drive toward disorder. When $\Delta S > 0$ the products are more disordered than the reactants.

The negative sign in front of the $T\Delta S$ shows that a positive ΔS makes a negative contribution to ΔG which tends to drive the reaction in the forward direction.

Note that increasing *T* increases the influence of ΔS on the reaction "driving force."

For a chemical reaction ΔH and ΔS are independent of each other. that is, you can not calculate one from the other. You can have situations where both are positive, both are negative, or one is positive and the other negative.

Notice that if ΔH and ΔS have the same sign they are working against each other. You can make the entropy win by increasing the temperature or you can make the enthalpy win by decreasing the temperature. We will show later that $\Delta_r G^o$ is related to the equilibrium constant for the reaction (as you might expect).

THIRD LAW OF THERMODYNAMICS

If we have sufficient heat capacity data (and the data on phase changes) we could write,

$$S(T) = S(T=0) + \int_0^T \frac{C_P}{T} dT. \qquad (1)$$

(If there is a phase change between 0 K and T we would have to add the entropy of the phase change.) If C_p were constant near $T = 0$ we would have,

$$S(T) = S(T=0) + C_p In\frac{T}{0},$$

which is undefined. Fortunately, experimentally $C_p \to 0$ as $T \to 0$. For nonmetals C_p is proportional to T^3 at low temperatures.

For metals C_p is proportional to T^3 at low temperatures but shifts over to being proportional to T at extremely low temperatures. (The latter happens when the atomic motion "freezes out" and the heat capacity is due to the motion of the conduction electrons in the metal.)

Equation 1 could be used to calculate absolute entropies for substances if we knew what the entropy is at absolute zero. Experimentally it appears that the entropy at absolute zero is the same for all substances. The third law of thermodynamics codifies this observation and sets

for all elements and compounds in their most stable and perfect crystalline state at absolute zero and one atmosphere pressure. (All except for helium, which is a liquid at the lowest observable temperatures at one atmosphere.)

The advantage of this law is that it allows us to use experimental data to compute the absolute entropy of a substance. For example, suppose we want to calculate the absolute entropy of liquid water at 25° C.

We would need to know the C_p of ice from 0 K to 273.15 K and the C_p of liquid water from 273.15 K to 298.15 K. We also need the heat of fusion of water at its normal melting point. With all of this data, which can be obtained partly from theory and partly from experiment, we find,

$$S^o_{H_2O}(25°C) = 0 + \int_0^{273.15} \frac{C_P(s)}{T} dT + \frac{\Delta H_{fus}}{273.15}$$

$$+ \int_{273.15}^{298.15} \frac{C_p(l)}{T} dT. \qquad (2)$$

Some substances may undergo several phase changes.

ENTROPY CHANGES IN CHEMICAL REACTIONS

We can use the third law entropies to calculate entropy changes for chemical reactions.

For a typical reaction,

$a\,\mathrm{A} + b\,\mathrm{B} \rightarrow c\,\mathrm{C} + d\,\mathrm{D}.$ (3)

the entropy change is

$\Delta_r S° = cS°_C + dcS°_D - acS°_A - bS°_B.$ (4)

Notice two things:

1. We did not define or use an entropy of formation, $\Delta_f S°$.
2. $S°_{element}$ is not zero.

As we have said before, $\Delta_r S°$ and $\Delta_r H°$ are independent of each other. They can not be calculated from each other. They must be calculated from Equation 4 and a comparable equation using heats of formation.

There is another way to calculate $\Delta_r S°$,

$$\Delta_r S° = \frac{\Delta_r H° - \Delta_r G°}{T} \qquad (5)$$

As we have seen before, the $\Delta_r G°$ and $\Delta_r H°$ can be calculated from free energies and heats of formation.

GIBBS-HELMHOLTZ EQUATION

Frequently we wish to run reactions at temperatures other than 25°C. Since we know that the change in the Gibbs free energy, $\Delta_r G°$, between products and reactants tells us whether or not the reaction will run spontaneously we will need this quantity at the new temperature.

(Reminder: If $\Delta_r G° < 0$ the reaction is spontaneous and if $\Delta_r G° > 0$ the reaction is not spontaneous.)

We also know that there are two components to $\Delta_r G^o$. That is, $\Delta_r G^o = \Delta_r H^o$ " $T\Delta_r S^o$, where the $\Delta_r H^o$ term is only weakly dependent on temperature, but the $T\Delta_r S^o$ term is strongly dependent on temperature due to the presence of the T in the term. So the question becomes, how do these things balance out? What is the dependence of $\Delta_r G^o$ on temperature? The simple answer is obtained from the derivative of G with respect to T.

$$\left(\frac{\partial G}{\partial T}\right)_p = -S. \tag{1}$$

If we apply Equation 1 to $\Delta_r G^o = G^o{}_{\text{products}} - G^o{}_{\text{reactants}}$ we get,

$$\left(\frac{\partial \Delta_r G^o}{\partial T}\right)_p = \left(\frac{\partial G^o{}_{product}}{\partial T}\right)_p - \left(\frac{\partial G^o{}_{reactants}}{\partial T}\right)_p$$

$$= -S^o_{products} - \left(-S^o_{reactants}\right)$$

$$= -\Delta_r S^o. \tag{2a, b, c}$$

Equation 2c shows that if $\Delta_r S^o$ is positive $\Delta_r G^o$ decreases with temperature, but if $\Delta_r S^o$ is negative $\Delta_r G^o$ increases with temperature. This will tell us whether $\Delta_r G^o$ increases or decreases with increasing temperature.

However, if we want to know whether or not a reaction is favored by an increase or decrease in temperature we really need to be looking at the equilibrium constant. If the equilibrium constant increases with temperature the reaction becomes more favored, but if the equilibrium constant decreases with temperature the reaction becomes less favored.

We will show later that the equilibrium constant for a chemical reaction depends on $\Delta_r G^o/T$ and not on $\Delta_r G^o$ all by itself. (There is a good reason for this which we will discuss below.) The Gibbs-Helmholtz equation addresses the question as to how $\Delta_r G^o/T$ changes with temperature. The Gibbs-Helmholtz equation can be presented in two different, but equivalent forms. If we are just worrying about G itself the two forms look like,

$$\left(\frac{\partial\left(\frac{G}{T}\right)}{\partial\left(\frac{1}{T}\right)}\right)_p = H, \tag{3a}$$

and

$$\left(\frac{\partial\left(\frac{G}{T}\right)}{\partial T}\right)_p = -\frac{H}{T^2}. \tag{3b}$$

Or, applying the procedure we used in Equations 2a, b, and c, we can write,

$$\left(\frac{\partial\left(\frac{\Delta G}{T}\right)}{\partial\left(\frac{1}{T}\right)}\right)_p = \Delta H, \tag{3c}$$

and

$$\left(\frac{\partial\left(\frac{\Delta G}{T}\right)}{\partial T}\right)_p = \frac{\Delta H}{T^2}. \tag{3d}$$

We won't derive any of these equations because the derivation is not particularly instructive. We will just show that the version of Equation 3a is true.

Start with the left-hand side of Equation 3a and show that it is equivalent to the definition of Gibbs free energy:

$$\left(\frac{\partial\left(\frac{G}{T}\right)}{\partial\left(\frac{1}{T}\right)}\right)_p = \left(\frac{\partial\left(\frac{G}{T}\right)}{\partial T}\right)_p \left(\frac{dT}{\partial\left(\frac{1}{T}\right)}\right)$$

$$= \left(\frac{\partial\left(\frac{G}{T}\right)}{\partial T}\right)_p \frac{1}{\frac{d\left(\frac{1}{T}\right)}{dT}}$$

$$= \left(\frac{\partial \left(\frac{G}{T} \right)}{\partial T} \right)_p \frac{1}{-\frac{1}{T^2}}$$

$$= -T^2 \left(\frac{\partial \left(\frac{G}{T} \right)}{\partial T} \right)_p$$

$$= -T^2 \left(+\frac{1}{T} \left(\frac{\partial G}{\partial T} \right)_p - \frac{1}{T^2} G \right)$$

$$= -T \left(\frac{\partial G}{\partial T} \right)_p + G$$

$$= -T(-S) + G$$

$$= G + TS$$

$$= H$$

We could easily substitute ΔG for G and end up with ΔH in the above sequence of equations.

So what is this good for? We will see that it is tremendously useful after we know the relationship between the equilibrium constant and $\Delta_r G°/T$, but for now let's use the Gibbs-Helmholtz equation to calculate $\Delta_r G°$ at a temperature, T_2 other than 25°C (which we will call T_1.

Set up Equation 3c for integration.

$$\left(\frac{\partial \left(\frac{\Delta_r G°}{T} \right)}{\partial \left(\frac{1}{T} \right)} \right)_p = \Delta_r H°, \tag{4}$$

$$d\left(\frac{\Delta_r G°}{T} \right) = \Delta_r H° d\left(\frac{1}{T} \right). \tag{5}$$

Equation 5 integrates to give,

$$\left(\frac{\Delta_r G°}{T} \right)_{T_2} - \left(\frac{\Delta_r G°}{T} \right)_{T_1} = \int_{T_1}^{T_2} \Delta_r H° d\left(\frac{1}{T} \right). \tag{6}$$

To carry out the integration on the right-hand side of

Equation 6 we would need to know how the heat of reaction changes with temperature. This information can be obtained if we know the heat capacities of the reactants and products as functions of temperature. However, usually the heat of reaction varies slowly with temperature so that it is a good approximation to regard the heat of reaction as constant.

Notice that Equation 5 is trivial to integrate if we make the approximation that $\Delta_r H^{\circ}$ is constant. With this approximation, Equation 5 integrates to,

$$\left(\frac{\Delta_r G^{\circ}}{T}\right)_{T_2} - \left(\frac{\Delta_r G^{\circ}}{T}\right)_{T_1} = \Delta_r H^{\circ} \int_{T_1}^{T_2} d\left(\frac{1}{T}\right), \qquad (7)$$

or,

$$\left(\frac{\Delta_r G^{\circ}}{T}\right)_{T_2} - \left(\frac{\Delta_r G^{\circ}}{T}\right)_{T_1} + \Delta_r H^{\circ}\left(\frac{1}{T} - \frac{1}{T^1}\right). \qquad (8)$$

We will apply Equation 8 to calculate ΔG for the freezing of super-cooled water at " 20°C.

The process is,

$$H_2O(l, -20^{\circ}C) \rightarrow H_2O(s, -20^{\circ}C).$$

Do we know ΔG for this process at 0°C? Note that freezing water at its normal melting point is a reversible process so that the heat of fusion is a reversible heat at constant temperature.

Therefore,

$$\Delta S_{fus} = \frac{\Delta H_{fus}}{T_{fus}}. \qquad (9)$$

Then,

$$\Delta G_{fus} = \Delta H_{fus} - T_{fus}\Delta S_{fus}$$

$$\Delta H_{fus} - T_{fus}\frac{\Delta H_{fus}}{T_{fus}}$$

$= 0.$ (10a, b, c)

Equation 8, for our problem, becomes,

$$\left(\frac{\Delta G}{T}\right)_{T_2} = \left(\frac{\Delta G}{T}\right)_{T_2} + \Delta H\left(\frac{1}{T_2} - \frac{1}{T_1}\right)$$

$$= 0 - \Delta H_{fus}\left(\frac{1}{T_2} - \frac{1}{T_1}\right)$$

$$\left(\frac{\Delta G}{253.15K}\right) = -6008J\left(\frac{1}{253.15K} - \frac{1}{273.15K}\right)$$

Then ΔG is – 440 J which is what we would expect because freezing super-cooled water is a spontaneous (and irreversible) process at " 20°C.

We said above that $\Delta_r G^o$ tells us whether or not a reaction wants to go, but that the equilibrium constant, the ultimate arbiter of how strongly the reaction wants to go, depends on $\Delta_r G^o/T$. We will prove this statement later, but for now it might be interesting to see why that might be. Consider any process at constant temperature and pressure. We know that if T and p are constant then,

$$\Delta G = \Delta H - T\Delta S. \quad (11)$$

Let's divide Equation 11 by T to see what $\Delta G/T$ looks like,

$$\left(\frac{\Delta G}{T}\right) = \left(\frac{\Delta H}{T}\right) - \Delta S. \quad (12)$$

Note that ΔH and ΔS in Equation 12 refer to things happening in the system. Rewrite Equation 12 to indicate that this is true,

$$\left(\frac{\Delta G}{T}\right) = \left(\frac{\Delta H_{sys.}}{T}\right) - \Delta S_{sys}. \quad (13)$$

Now, ΔH_{sys} is heat absorbed by the system. Where did this heat come from? It had to come from the surroundings so.

$$\Delta H_{sun} = -\Delta H_{sys}$$

$$\Delta H_{sys} = -\Delta H_{sun}. \quad (14a, b)$$

Most likely our process is irreversibly, but that doesn't matter because H is a state function so that ΔH is independent of path. We can always find a reversibly path to change the enthalpy of the surroundings by an amount ΔH_{surr}. (Our process is at constant temperature and pressure so that ΔH in both cases is a heat at constant pressure.) If we add heat ΔH_{surr} to the surroundings isothermally and reversibly then the entropy change in the surroundings is,

$$\Delta S_{sun} = \frac{\Delta H_{sun}}{T}$$

$$= -\frac{\Delta H_{sys}}{T}, \tag{15a, b}$$

or

$$\frac{\Delta H_{sys}}{T} = -\Delta S_{sun}. \tag{16}$$

Applying the result of Equation 16 to Equation 13 we find that,

$$\frac{\Delta G}{T} = -\Delta S_{sun} - \Delta S_{sys}$$

$$= -\Delta S_{universe}. \tag{17a, b}$$

Equation 17b tells us several things. It tells us that, in the final analysis, the ultimate driving force in nature is entropy, that is, the drive toward disorder. The system plus the surroundings is a closed isolated system so that the only spontaneous processes allowed are those which increase the entropy. Secondly, it explains why $\Delta G/T$ is more important than simply ΔG in determining how strongly spontaneous a process is. It is "$\Delta G/T$ which is related to the entropy change of the universe, not ΔG directly. Equation 17b also shows that the Gibbs free energy manages to include the entropy effects in the surroundings without ever telling us that it is doing so.

GIBBS FREE ENERGY AND PRESSURE

The Gibbs free energy depends on pressure as well as on temperature. The pressure dependence of the Gibbs free energy in a closed system is given by the combined first and second laws and the definition of Gibbs free energy as,

$$dG = sdT + Vdp \tag{1}$$

If we hold temperature constant and vary only the pressure we can prepare Equation 1 for integration from pressure p_1 to p_2 as follows:

$$dG = Vdp, \tag{2}$$

Then

$$G(p_2) - G(p_1) = \int_{p_1}^{p_2} \bar{V} dp \tag{3}$$

or

$$G(p_2) = G(p_1) + \int_{p1}^{p2} Vdp \quad (4)$$

Equation 4 is quite general and applies to all isotropic substance: solids, liquids, ideal gases, and real gases. We will apply it first to isotropic solids and liquids.

SOLIDS AND LIQUIDS

We know that solids and liquids are not very compressible so, to a first approximation, we can regard the volume in Equation 4 as constant (as long as the range of pressure is not too large). Then the *V* in Equation 4 comes out of the integral and we can integrate easily to get,

$$G(p_2) = G(p_1) + V(p_2 - p_1) \quad (5)$$

We know, however, that solids and liquids are slightly compressible and that we can define the isothermal compressibility as

$$k = -\frac{1}{V}\left(\frac{\partial V}{\partial p}\right)_T. \quad (6)$$

Our second level of approximation is to regard *k* as approximately constant. (In fact, *k* is not constant, but the variation with pressure is so small it can be ignored unless enormous pressures - megabars - are involved.) With *k* regarded as constant we can rearrange Equation 6 and integrate it to find an expression for *V* as a function of *p* (which can then be substituted into Equation 4 and integrated.) Rearrangement of Equation 6 looks like,

$$\frac{dV}{V} = -kdp \quad (7)$$

Integrate from p_1 to p_2 (and volume goes from V_1 to V_2) to get,

$$In\frac{V_2}{V_1} = -k(p_2 - p_1) \quad (8)$$

Take the antilog of both sides,

$$V_2 - V_{1e}^{-k(p_1 - p_2)} \quad (9)$$

In Equation 9 we can let p_1 be a constant and let p_2 range over the pressures of interest. There is no reason why we have to keep the subscript "2" on p_2 so change p_2 to just p.

This gives us V as a function of p,

$$V_2 - V_{1e}^{-k(p-p_1)=V_{1e}^{kp_1}e^{-kp}}. \tag{10}$$

On the far right of Equation 10 the constant parts have been separated from the variable part to make it easy to integrate. When this expression for V is plugged into Equation 4 only the e^{-kp} need stay inside the integral.

IDEAL GASES

Equation 4 is also valid for gases, only here we put in the value of V for an ideal gas.

$$V = \frac{nRT}{p} \tag{11}$$

With this substitution Equation 4 becomes,

$$G(p_2) = G(p_1) + \int_{p_1}^{p_2} \frac{nRT}{p} dp \tag{12}$$

which is an integral we have done many times. After integration Equation 12 becomes.

$$G(p_2) = G(p_1) + nRT\, In \frac{p_2}{p_1} \tag{13}$$

It is customary (and useful) to make several changes in Equation 13. We let p_2 range over the pressures of interest to us and call it just p, we let p_1 be some standard state pressure and call it p^o, and finally we divide through by the number of moles of gas, n.

With these changes equation 13 is written,

$$\frac{G(p)}{n} = \frac{G(p^\circ)}{n} + RT\, In \frac{p}{p^\circ} \tag{14}$$

One more change: the quantity G/n turns out to be so important that it is given a special symbol and its own name. Strictly speaking G/n is just the Gibbs free energy per mole of substance, but the simplicity belies its importance.

This quantity is called the chemical potential and it is

given the symbol, μ. Our final version of what used to be Equation 13 is now,

$$\mu = \mu^\circ + RT\, In \frac{p}{p^\circ} \tag{15}$$

We have replaced $G(p^o)/n$, the molar Gibbs free energy at the standard state pressure, with its chemical potential symbol, μ^o. In most cases we will set the standard state pressure equal to one atmosphere. It is not unusual to see Equation 15 written,

$$\mu = \mu^\circ + RT\, In\, p, \tag{16}$$

but when it is written like this we have to remember that there is an implied $p^o = 1$ atm dividing the p in the ln p, otherwise the argument of the log function would not be unitless.

NONIDEAL GASES

Equation 15 was derived assuming the gas is ideal. It does not apply to real gases or approximations to a real gas, like the van der Waals equation of state. If we know the equation of state we can go back to Equation 4 and make the appropriate modifications in our notation. That is, we divide Equation 4 by the number of moles, n, let p_1 equal the standard state pressure, p^o and note that V/n is the molar volume to get,

$$\mu = \mu^\circ + \int_{p}^{p} \cdot \overline{V}(p)\, dp' \tag{17}$$

We have also let p_2 range over the pressures of interest and called it just plain p, which means that we have to change the dummy variable of integration from p to p'.) Equation 17 would provide the correct answer in numerical calculations, but it would wreak havoc in some of the later developments of thermodynamics, namely the equilibrium constant expression, as we will see later. G. N Lewis (the same Lewis of the Lewis dot structures and Lewis acid/base theory) proposed to preserve the form of Equation 15 by writing the chemical potential as,

$$\mu = \mu^\circ + RT\, In \frac{f(p)}{p^\circ} \tag{18}$$

This equation defines a quantity $f(p)$ called the fugacity.

The fugacity has units of pressure and it is a function of pressure. It contains all the information on the nonideality of the gas. For an ideal gas the fugacity is the same as the pressure. Since all real gases become ideal in the limit as pressure goes to zero we must have.

$$\lim_{p \to 0} f(p) = p \tag{19}$$

We would like to have a way of calculating the fugacity from the equation of state for a gas. To do this go all the way back to Equation 2 and divide it by the number of moles, n,

$$d\frac{G}{n} = \frac{V}{n}dp, \tag{20}$$

or, in our new notation,

$$d\mu = \overline{V}dp. \tag{21}$$

We can get another expression for $d\mu$ by taking the differential of Equation 18 (Remember that R, T, p^o and μ^o are constants.)

$$d\mu = RTd\ln f. \tag{22}$$

The $d\mu$ in Equations 21 and 22 must be the same, so we can set them equal to each other

$$RTd\ln f = \overline{V}dp. \tag{23}$$

Rearrange this to get,

$$d\ln f = \frac{\overline{V}}{RT}dp \tag{24}$$

We could integrate this equation directly, but that would sort of take us back to where we started from. Instead, we use a mathematical trick before we integrate it. Add and $\frac{dp}{p}$ subtract to the right hand side of Equation 24,

$$d\ln f = \frac{\overline{V}}{RT}dp - \frac{dp}{p} + \frac{dp}{p}. \tag{25}$$

You can see that we didn't really change anything. Regroup the terms in Equation 25,

$$d\ln f = \left(\frac{\overline{V}}{RT} - \frac{1}{p}\right)dp + \frac{dp}{p}. \tag{26}$$

Now integrate from p^o to p. (We will have to call our dummy variable of integration p' so as not to conflict with the limits of integration.) We get,

$$In\, f - In\, f_o = \int_{p_o}^{p}\left(\frac{\overline{V}(p)}{RT} - \frac{1}{p'}\right)dp' + In\, p - In\, p_o \qquad (27)$$

where f_o is the fugacity at p_o. Move the to the right hand side,

$$In\, f = \int_{p_o}^{p}\left(\frac{\overline{V}(p)}{RT} - \frac{1}{p'}\right)dp' + In\, p\left(In\, f_o - In\, p_o\right) \qquad (28)$$

Now we can take the limit where p_o goes to zero. We know that f_o goes to p_o as $p_o \to 0$ so the last two terms in parentheses on the right cancel each other in this limit. Equation 28 becomes,

$$In\, f = \int_{0}^{p}\left(\frac{\overline{V}(p)}{RT} - \frac{1}{p'}\right)dp' + In\, p \qquad (29)$$

Equation 29 will suffice to calculate the fugacity, but it is customary to take the antilog of both sides to get,

$$f = pe^{\int_0^p}\left(\frac{\overline{V}(p)}{RT} - \frac{1}{p'}\right)dp' + pe^{\int_0^p \frac{1}{p'}(Z-1)dp'} \qquad (30)$$

In the last segment of Equation 29, $z = \frac{p'\overline{V}}{RT}$, is the so-called compressibility factor. In either version of Equation of 29 it is easy to see that if the gas is ideal $f = p$. It requires an equation of state or experimental data to calculate a fugacity from either Equation 30 or Equaton 29. From the right-hand side of Equation 30 we can see that the second form of the virial expansion,

$$\frac{p\overline{V}}{RT} = 1 + B'p + C'p^2 + ... \qquad (31)$$

would be the best choice for calculating fugacity

OPEN SYSTEMS

Up to now all of the systems we have been working with were closed systems. That is, no material moved in or out of the system. Now it is time to extend our discussion to open systems, in which material *can* move in or out of the system.

As usual we begin with the combined first and second

laws of thermodynamics, only now we have to take into account that the internal energy, U, will depend on the number of moles of each component present.

We write,

$$U = U(S, V, n_1, n_2, n_3, ...), \tag{1}$$

so, instead of

$$dU = TdS\text{-}pdV \tag{2}$$

we must add the contribution of moving material in and out of the system.

That is, we write,

$$dU = \left(\frac{\partial U}{\partial S}\right)_{V, n_1 n_2, ...} dS + \left(\frac{\partial U}{\partial V}\right)_{S, n_1 n_2, ...} dV$$

$$+\left(\frac{\partial U}{\partial n_1}\right)_{S, V, n_2 n_3, ...} dn_1 + \left(\frac{\partial U}{\partial n_2}\right)_{S, V, n_1 n_3, ...} dn_2 + ... \tag{3}$$

The first two terms of Equation 3 must be the same as the combined first and second laws, Equation 2. The remaining terms in Equation 3 are new to us. They are clearly awkward to write so we invent a new symbol for the partial derivative,

$$\left(\frac{\partial U}{\partial n_1}\right)_{S, V, n_2 n_3 ....} = \mu_1,$$

$$\left(\frac{\partial U}{\partial n_2}\right)_{S, V, n_2 n_3 ....} = \mu_2, \tag{4a, b}$$

and so on.

Notice that we are using the same symbol, μ, here that we used for G/n previously. That is not an accident, as we shall see. We called μ the chemical potential. In an open system with more than one component μ_i will be the chemical potential of component i in the mixture.

With this notation we can rewrite Equation 3 as,

$$dU = TdS - pdV + \mu_1 dn_1 + \mu_2 dn_2 + \mu_3 dn_{3+} ...; \tag{5}$$

We can carry the terms accounting for the movement of material through to our other "energy" functions, H, A, and G.

From $H = U + pV$ we obtain,

$$dU = TdS - Vdp + \mu_1 dn_1 + \mu_2 dn_2 + \mu_3 dn_{3+} ...; \tag{6a}$$

from $A = U - TS$ we obtain

$$dA = -SdT - pdV + \mu_1 dn_1 + \mu_2 dn_2 + \mu_3 dn_{3+}...; \quad (6b)$$

and from $G = U + pV$ " TS we obtain,

$$dG = -SdT + Vdp + \mu_1 dn_1 + \mu_2 dn_2 + \mu_3 dn_{3+}.... \quad (6c)$$

From Equations 5 and 6a, b, and c we see that there are four different appearing ways to write the μ_i's.

For example,

$$\mu_1 = \left(\frac{\partial U}{\partial n_1}\right)_{S,V,n_1n_2,....} = \left(\frac{\partial H}{\partial n_1}\right)_{S,p,n_1n_2,...} = \left(\frac{\partial A}{\partial n_1}\right)_{T,V,n_2n_3,...}$$

$$= \left(\frac{\partial G}{\partial n_1}\right)_{T,V,n_1n_2,...} \quad (7)$$

All of these definitions are equivalent, but the last one,

$$\mu_1 = \left(\frac{\partial G}{\partial n_1}\right)_{T,p,n_1n_2,....}, \quad (8)$$

will be the most important one because it gives the change in Gibbs free energy that comes from adding or removing material at constant pressure and temperature. For a process a constant temperature and pressure dG becomes

$$dG = \mu_1 dn_1 + \mu_2 dn_2 + \mu_1 dn_1 + ...$$

$$\sum_i \mu_i dn_i \quad (9a, b)$$

Notice that the quantities, μ_i, are intensive variables. We can see this because they are an extensive variable, G divided by another extensive variable, n_i.

INTEGRATION OF d*U*

There are some unique features of the differential, dU, in Equation 5,

$$dU = TdS - pdV + \mu_1 dn_1 + \mu_2 dn_2 + \mu_3 dn_3 + ..., \quad (5)$$

which allow us to integrate it in an exceptionally simple manner. (This statement is not true for the differentials dH, dA, and dG.) In Equation 5 all the differentials (dU, dS, dV, dn_1, etc.) are differentials of extensive properties. That is, they all depend on the amount of material in the system. In addition, all the coefficients of these differentials (T, p, μ_1, etc.) are intensive properties. We can imagine integrating Equation 5 by starting out with the system in one container and transferring it to another container one differential drop at a time while holding all of the intensive variables constant. As we move the system from one container to another all the extensive quantities move to the new container in proportion to the size of the drop. Let's parameterize this process with a variable, x,

$$0 \leq x \leq 1,$$

where $x = 0$ means the system is entirely in its old container and $x = 1$ means the system has been entirely transferred to its new container. The "drop size" is given by the size of dx. Then,

$$dU = U\,dx,$$
$$dS = S\,dx,$$
$$dV = V\,dx,$$
$$dn_1 = n_1\,dx,$$
$$dn_2 = n_2\,dx, \tag{10}$$

and so on.

We can now rewrite Equation 5 in terms of the differential, dx,

$$Udx = TSdx - pVdx + \mu_1 n_1 dx + \mu_2 n_2 dx + \mu_3 n_3 dx + ..., \quad \text{(11a, b)}$$

In actual fact we don't have to integrate Equation 11a or 11b to get the desired result, just divide 11b by dx to get

$$Udx = (TS - pV + \mu_1 n_1 + \mu_2 n_2 + \mu_3 n_3 + ...)dx. \tag{12}$$

If you insist on moving the system from one container to the other then we can do this by integrating Equation 11b from $x = 0$ to $x = 1$,

$$U\int_0^1 dx = TS - pV + \mu_1 n_1 + \mu_2 n_2 + \mu_3 n_3 + ...)\int_0^1 dx, \tag{13}$$

which gives Equation 12 because the integral is equal to 1.

Equation 12 is called the integrated form of the combined first and second laws for an open system.

We can use Equation 12 to obtain "integrated forms" of H, A, and G,

$$H = U + pV = TS + \mu_1 n_1 + \mu_2 n_2 + \mu_3 n_3 + ...,$$
$$A = U - TS = -pV + \mu_1 n_1 + \mu_2 n_2 + \mu_3 n_3 + ..., \quad \text{(14a, b)}$$

but the most important one is,

$$G = U - pV - TS = \mu_1 n_1 + \mu_2 n_2 + \mu_3 n_3 + ..., \quad (14c)$$

The latter equation is sometimes written,

$$G = \sum_i \mu_i n_i, \quad (15)$$

where the summation is over all the components in the system.

THE GIBBS-DUHEM EQUATION

We now have (or can get) two different expressions for dG. One expression is Equation 6c,

$$dG = SdT + Vdp + \mu_1 dn_1 + \mu_2 dn_2 + \mu_3 dn_3 + ..., \quad (6c)$$

and the other can be obtained from Equation 15 as,

$$dG = \mu_1 dn_1 + \mu_2 dn_2 + \mu_3 dn_3 + ... n_1 d\mu_1 + n_2 d\mu_2 + n_3 d\mu_3 + ... \quad (16)$$

Setting Equations 6c and 16 equal to each other, and canceling terms that are the same on both sides we obtain,

$$-SdT + Vdp = n_1 d\mu_1 + n_2 d\mu_2 + n_3 d\mu_3 + ... \quad (17)$$

Equation 17 is called the Gibbs-Duhem equation. It tells us that the intensive variables in a system can not all be assigned values independently. That is, you can assign virtually any value you desire to all of the intensive variables but one, but the value of that last one will be predetermined by the values of the others.

The Gibbs-Duhem equation is most often used for processes at constant temperature and pressure, whence Equation 17 becomes,

$$n_1 d\mu_1 + n_2 d\mu_2 + n_3 d\mu_3 + ... = 0. \quad (18)$$

Consider a two-component system at constant T and p. Then,

$$n_1 d\mu_1 + n_2 d\mu_2 = 0, \quad (19a)$$

or

$$n_1 d\mu_1 = -n_2 d\mu_2, \quad (19b)$$

which tells us that if we change μ_2 then μ_1 responds with a change which depends on the ratio of the amounts of the two components.

In a one-component system the Gibbs-Duhem equation takes the form,

$$-SdT + Vdp = n_1 d\mu_1. \quad (20)$$

We rearrange this equation to give,

$$d\mu_1 = -\frac{S}{n_1}dT + \frac{V}{n_1}dp$$

$$\bar{S}dT + \bar{V}dp, \qquad (21a, b)$$

which demonstrates that in a one component system the chemical potential is a function only of T and p.

COMMENT ON LEGENDRE TRANSFORMS

Recall that we get from Equation 12 to Equations 14a, b, and c by making Legendre transforms. Equation 14c shows that G is a natural function of T, p, and all the n_i's. Can we now make more Legendre transforms to obtain a new function which is a function of T, p, and all the μ_i's as in,

$$New\ Function = U - TS + pV - \mu_1 n_1 - \mu_2 n_2 - \mu_3 n_3 - ....? \qquad (22)$$

If we plug in Equation 12 for U here we get,

New Function = 0

This means that no such function exists. This is a consequence of the fact that the new function, if it existed, would be a function of only intensive variables (all independent) and we have seen from the Gibbs-Duhem equation that the intensive variables are not all independent.

MAXWELL'S RELATIONS REVISITED

Equations 5 and 6a, b, and c open up many new possibilities for Maxwell's relations. For example, from Equation 6c we obtain two equations which may be useful later,

$$\left(\frac{\partial S}{\partial n_1}\right)_{T,p} = -\left(\frac{\partial \mu_1}{\partial T}\right)_{p,n_1}, \qquad (23)$$

and

$$\left(\frac{\partial S}{\partial n_1}\right)_{T,p} = -\left(\frac{\partial \mu_1}{\partial T}\right)_{T,n_1}. \qquad (24)$$

In equations 23 and 24 it is understood that all the other n's are also being held constant. Equations 23 and 24 will be useful when we start worrying about how the chemical potentials depend on temperature and pressure.

PHASE EQUILIBRIUM

We are all familiar with phase transitions where a substance transforms from one stable phase to another at an equilibrium temperature. For example, ice will be in equilibrium with liquid water at 273.15 K and 1 atmosphere pressure, or liquid water will be in equilibrium with water vapour at 373.15 K and one atmosphere. We will now take a look at the thermodynamic principles involved in phase transitions. I will assume at the beginning that we know nothing about the conditions under which two phases can be in equilibrium. The only things we know are the criteria for equilibrium under certain conditions. That is, we know that,

$$dS_{U,V} \geq 0 \tag{1}$$

for a closed isolated system (that is, the entropy seeks a maximum), and

$$dA_{T,V} \leq 0 \tag{2}$$

for a system as constant temperature and volume (that is,the Helmholtz free energy seeks a minimum), and

$$dG_{T,p} \leq 0 \tag{3}$$

for a system at constant temperature and pressure (the Gibbs free energy seeks a minimum). We will do a series of three thought experiments under different sets of conditions and use the above criteria to tell us things about the temperature, pressure, and chemical potential of a system of two phases in equilibrium.

I. A closed isolated system.

Consider a closed isolated system consisting of two phases, phase α in equilibrium with phase β. We will call the temperature of the α phase T_α and the temperature of the β phase T_β.

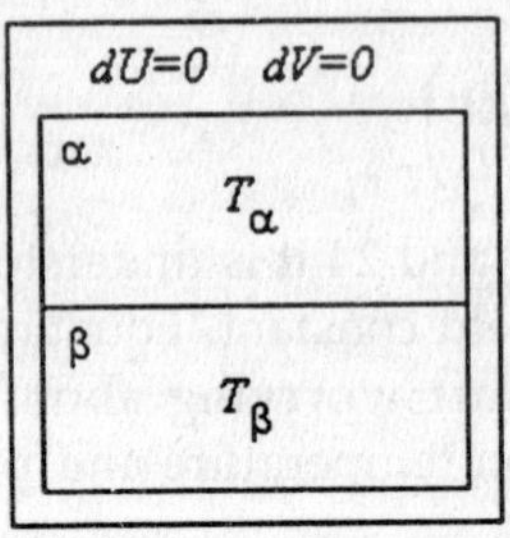

We do not know, yet, what is the relationship between T_α and T_β. We do know that in this system entropy seeks a maximum.

That is,

$$dS_{U,V} \geq 0. \tag{4}$$

Let us transfer a small amount of heat, *dq*, reversibly from phase α to phase β. For definiteness we will set $dq > 0$.

So,

$$dq_\alpha = -dq$$
$$dq_\beta = +dq, \tag{5a,b}$$

and

$$dS_\alpha = \frac{dq_\alpha}{T_\alpha} = \frac{-dq_\alpha}{T_\alpha},$$
$$dS_\beta = \frac{dq_\beta}{T_\beta} = \frac{-dq_\beta}{T_\beta}, \tag{6a, b}$$

The total entropy change for the system is,

$$dS = dS_\alpha + dS_\beta = \frac{dq_\alpha}{T_\alpha} + \frac{dq_\beta}{T_\beta} = \left(\frac{1}{T_\beta} - \frac{1}{T_\alpha}\right) dq \geq 0. \tag{7}$$

Since *dq* is positive by construction we conclude that,

$$\left(\frac{1}{T_\beta} - \frac{1}{T_\alpha}\right) \geq 0,$$
$$\left(\frac{1}{T_\beta} \geq \frac{1}{T_\alpha}\right), \tag{8}$$

or

$$T_\beta \leq T_\alpha \tag{9}$$

If the system is not at equilibrium then $T_\beta < T_\alpha$, which makes sense because heat flows spontaneously from a higher temperature to a lower temperature.

If the system is at equilibrium then $T_\beta = T_\alpha = T$. There is only one temperature defined and there in no need to distinguish between the temperatures of the two phases.

II. A system at constant volume and temperature

Since we now know that at equilibrium there is only one temperature defined for the two phases let us now remove the system from isolation and place it in a heat bath at temperature, T. We will still hold the volume constant. Under conditions of constant temperature and volume we know that the criterion for equilibrium is that the Helmholtz free energy seeks a minimum.

That is,

$dA_{T,V} \leq 0.$ Our system now looks like,

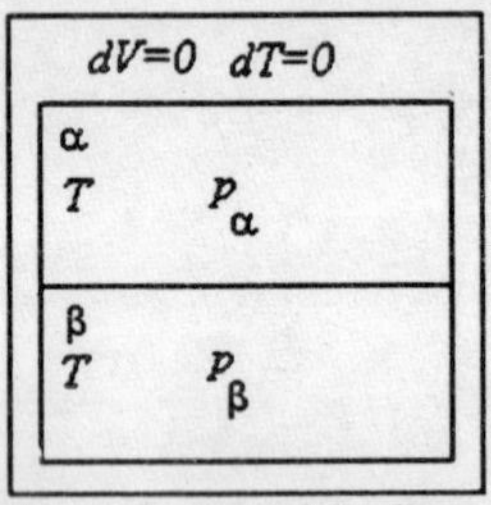

but we do not know the relationship between the pressures in the two phases.

Let us now transfer a small amount of volume, $dV > 0$, from phase α to phase β.

We know that dA = " SdT " pdV = " pdV for a constant temperature system.

So,

$$dA_\alpha = p_\alpha dV_\alpha = -p_\alpha(-V),$$

$$dA_\beta = p_\beta dV_\beta = -p_\beta dV. \qquad (10)$$

The change in Helmholtz free energy for the entire system is,

$$dA = dA_\alpha + dA_\beta = p_\alpha dV - p_\beta dV = (p_\alpha - p_\beta)dV \leq 0. \qquad (11)$$

Since dV is positive we conclude that,

$$p_\alpha - p_\beta \leq 0, \qquad (12)$$

or

$$p_\alpha \leq p_\beta.$$

If the system is not at equilibrium then $p_\alpha \leq p_\beta$, which makes sense since the β phase expanded at the expense of the α phase. If the system is at equilibrium then $p_\alpha = p_\beta = p$, so that there is only one pressure defined for the system.

III. The system at constant pressure and temperature

Since we now know that at equilibrium both phases are at the same temperature and pressure, let us remove the constant volume restriction and place our system in a heat bath at temperature, T, and a pressure bath at pressure, p. (The atmosphere is a good example of a pressure bath at approximately one atmosphere pressure. You can make chambers to hold different pressures if you wish.) Our system looks like,

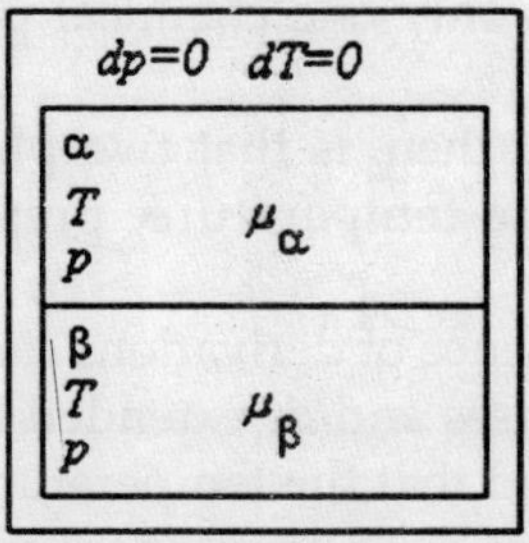

but we do not know the relationship between the chemical potentials in the two phases.

At constant temperature and pressure we know that

$$dG_{T,P} \leq 0. \tag{13}$$

We also know that,

$$dG = -SdT + Vdp + \mu\, dn,$$

which at constant temperature and pressure becomes,

$$dG_{T,P} = \mu\, dn. \tag{14}$$

Let us now move a small amount of material, dn, from phase α to phase β (with $dn > 0$). Then,

$$dn_\alpha = -dn,$$
$$dn_\beta = +dn. \tag{15a, b}$$

and

$$dG_\alpha = \mu_\alpha\, dn_\alpha = -\mu_\alpha\, dn,$$
$$dG_\beta = \mu_\beta\, dn_\beta = -\mu_\beta\, dn, \tag{16a, b}$$

The change in Gibbs free energy for the entire system is then

$$dG = dG_\alpha + dG_\beta = \mu_\alpha\, dn + \mu_\beta\, dn = \left(\mu_\beta - \mu_\alpha\right) dn. \tag{17}$$

Since dn is positive by construction we conclude that,

$$\mu_\beta - \mu_\alpha \le 0, \quad (18)$$

or

$$\mu_\beta \le \mu_\alpha \quad (19)$$

If the system is not in equilibrium then which makes sense since material wants to move from a region of higher chemical potential to a region of lower chemical potential.

If the system is in equilibrium then,

$$\mu_\beta = \mu_\alpha = \mu, \quad (20)$$

so that there is only one chemical potential defined for the system.

Our conclusion, then, is that two phases in equilibrium must have the same temperature, pressure and chemical potential.

The above sequence of derivations can easily be extended to include more phases and/or extended to include mixtures, where we would find that the temperature, pressure, and the chemical potential of each component must be the same in every phase. We will use these facts a little later in our derivation of the Gibbs phase rule.

ONE-COMPONENT PHASE DIAGRAMS

We know that most substances can exist in more than one phase. For example, solid, liquid, and gas. Some substances even have more than one solid phase.

Carbon exists as graphite, diamond and a whole encyclopedia of forms related to C_{60}; tin exists as white tin and gray tin; and so on. A great amount of information on the phases of a substance can be summarized on one diagram called a "phase diagram."

Imagine that we take a sheet of graph paper and label the vertical axis (the ordinate) "*p*" for pressure, and the horizontal axis (the abscissa) "*T*" for temperature. Now examine the system at a particular temperature and pressure and label the corresponding point on the graph paper with a letter indicating the particular phase the system is in. That is, label the point with a "g," if the system is a gas, "l," for a liquid, "s," for a solid, and so on.

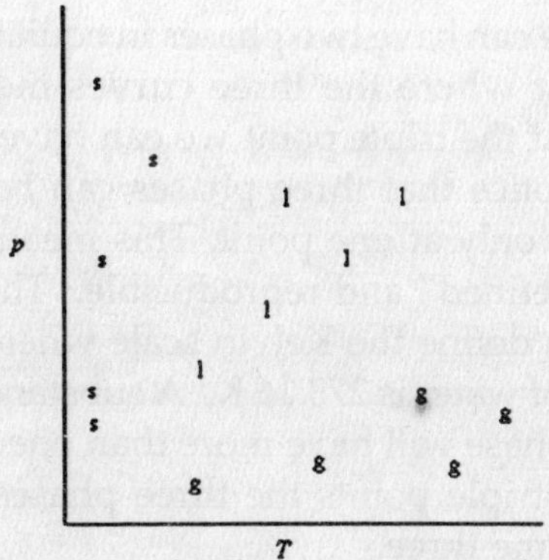

If we continued to do this at more and more points we would find that the graph paper would be separated into regions. In one of the regions all the points would be a "g" indicting that the substance is gas at those temperatures and pressures. In another region all the points would correspond to the substance being a liquid, and so on. In the limit where we considered an infinite number of points the regions would be separated from each other by curves.

Once we know the phase of a system in a particular region it is not usual to use so many g's and l's and s's. Most often we just label each region once.

This diagram is an example of a phase diagram. The line separating the solid and liquid regions is called the "melting curve," the line separating the liquid and gas regions is called the "vapourization curve," or the "vapour pressure curve," and the line separating the solid and gas regions is called the "sublimation curve." In the regions between the lines only one phase, the phase indicated, can exist at equilibrium.

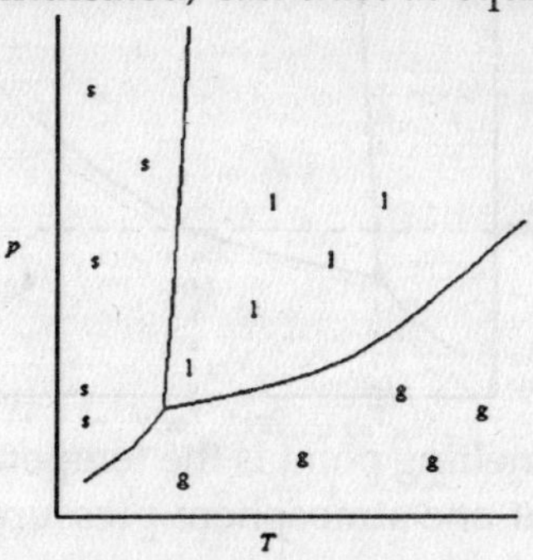

On a line we can have two phases in equilibrium with each other. The point where the three curves meet is called the "triple point." At the triple point we can have three phases in equilibrium. Notice that three phases can be at equilibrium with each other only at one point. This means that the triple point is "well defined" and reproducible. The triple point of water is used to define the Kelvin scale where, by definition, the triple point of water is 273.16 K. A substance that has more than one solid phase will have more than one triple point, but at the different triple points the three phases in equilibrium are never the same three.

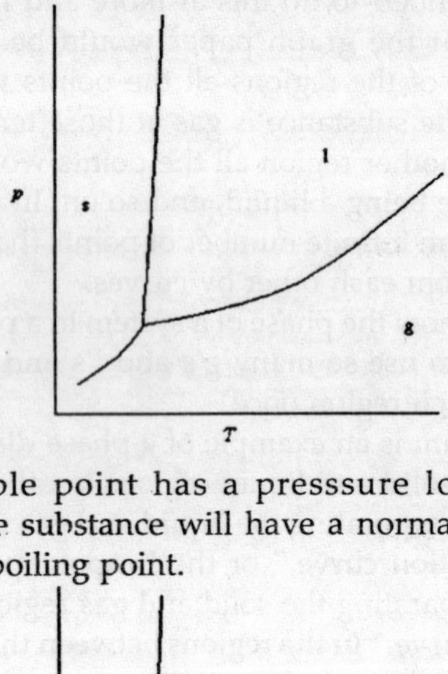

If the triple point has a presssure lower than one atmosphere the substance will have a normal melting point and a normal boiling point.

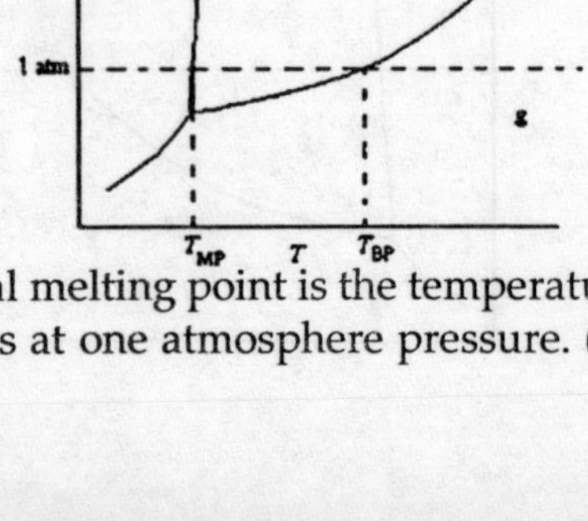

The normal melting point is the temperature at which the substance melts at one atmosphere pressure. (As you can see the melting point will be slightly different at other pressures.) The normal boiling point is the temperature where the substance boils at one atmosphere pressure. Another way to state this is that at the normal boiling point the vapour pressure of the substance equals one atmosphere.

As we learned in the early sections of the course, there is a point, called the critical point, "above" which the substance can not be converted to a liquid. That means that the vapour pressure curve must end at the critical point.

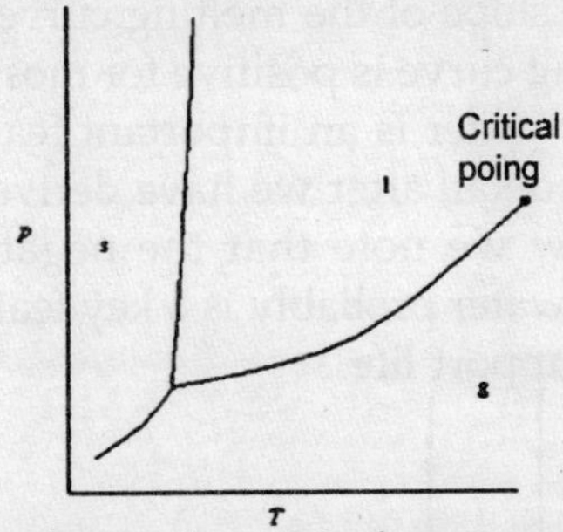

At pressures above the critical pressures it usually does not make sense to try to distinguish between whether the substance is a solid or a liquid. In these cases the substances is referred to simply as a "fluid."

CO_2 is a substance that does not have a normal melting point or boiling point because the substance is not liquid at one atmosphere pressure. (Its triple point pressure is above one atmosphere.) A phase diagram for CO_2 might look like the ones we have been showing, but the one-atmosphere line would be below the triple point.

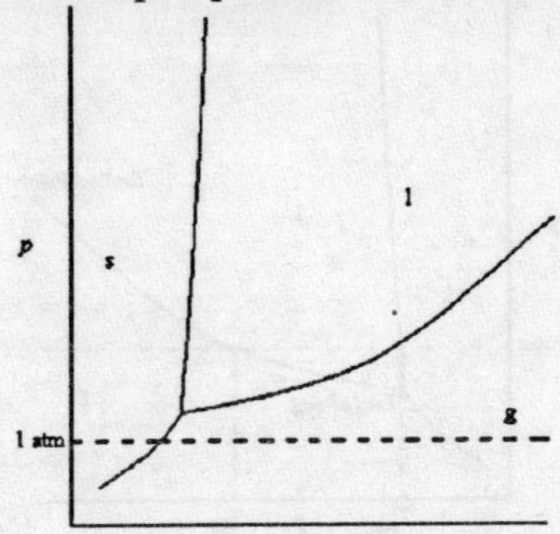

Solid CO_2 is called "dry ice" because it never melts at pressures near one atmosphere. At one atmosphere pressure the solid "sublimes" directly from the solid into the gas phase. In a CO_2 fire extinguisher you can hear the liquid "slosh around" inside the tank because liquid CO_2 does exist at pressures above the triple point pressure. Water is an unusual substance because its melting point (273.15 K) is at a lower temperature than the triple point temperature (273.16 K). A schematic phase diagram for water might look something like,

Note that the slope of the melting curve is negative. The slope of the melting curve is positive for most substances. The negative slope for water is an important feature and we will discuss it in more detail after we have derived the Clapeyron equation. For now we note that the negative slope of the melting curve for water probably is a key feature in the ability of our planet to support life.

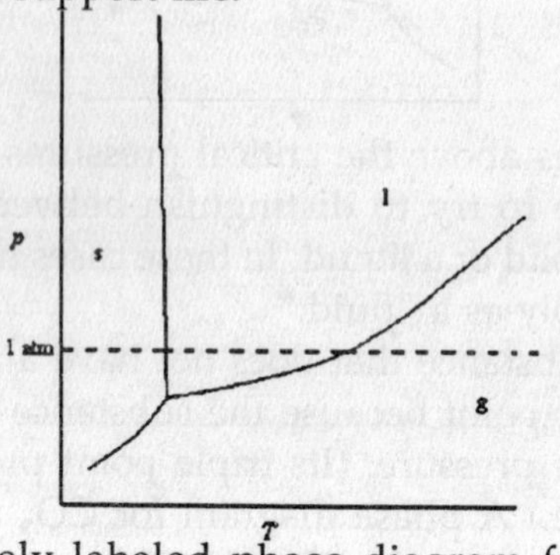

A completely labeled phase diagram for an ordinary substance might look like this,

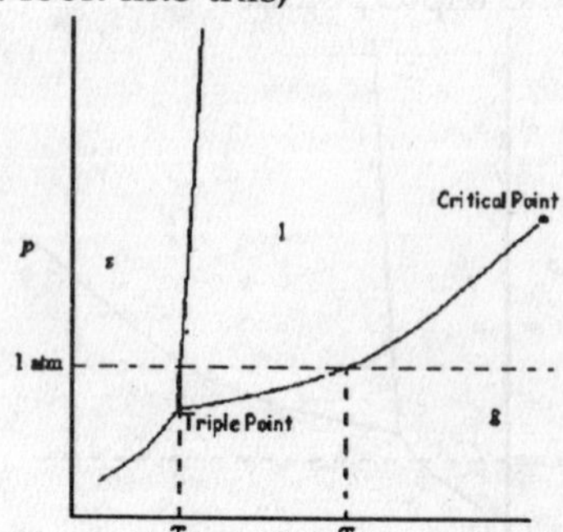

The dashed lines have been added simply to help locate the normal melting and boiling points of the substance and are not a necessary part of the phase diagram.

CLAPEYRON EQUATION

In a one-component phase diagram the lines separating the phases (for example, solid from liquid, liquid from vapour, and so on) are not arbitrary, they are determined by the principles of thermodynamics and equilibrium. Thermodynamics provides an equation for each line which gives pressure, p, as a function of temperature T, in other words an equation of the form $p = p(T)$. It is now our task to find the form of the function $p(T)$. The diagram below is a portion of a hypothetical phase diagram and the phase line on this diagram separates the phases α and β. In other words, the phases α and β are in equilibrium with each other at any and all points on the line.

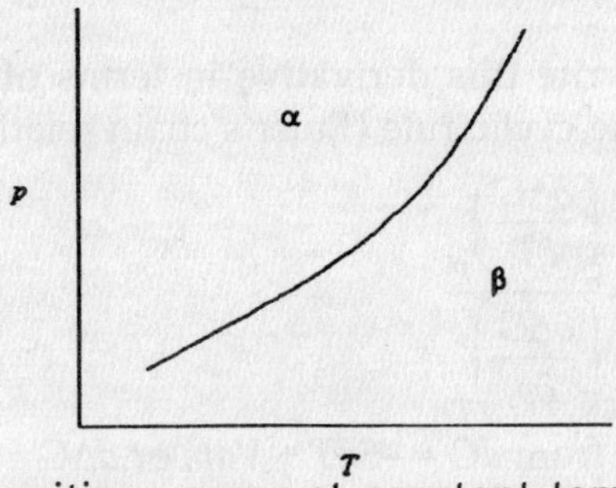

Phase transitions occur at constant temperature and pressure at the points on this line. Consider the phase transition,

$$\alpha \rightarrow \beta. \quad (1)$$

Since the α and β phases are in equilibrium with each other at any point on the line we know that the change in the Gibbs free energy for the transition given by Equation 1 is zero everywhere on the α-β phase line. That is,

$$\Delta G = 0. \quad (2)$$

We already know that for any process at constant temperature it must be true that

$$\Delta G = \Delta H - T\Delta S, \quad (3)$$

so that,

$$\Delta S = \frac{\Delta H}{T} \tag{4}$$

for a phase transition.

The various quantities appearing in these equations are defined as usual by,

$$\Delta G = G_\beta - G_\alpha \tag{5}$$

$$\Delta S = S_\beta - S_\alpha \tag{6}$$

$$\Delta H = H_\beta - H_\alpha \tag{7}$$

We will also define ΔV the same way,

$$\Delta V = V_\beta - V_\alpha \tag{8}$$

As we have already said, $\Delta G = 0$ along the α-β phase line. That is, ΔG is constant along the line and it is being held constant at the particular value of zero. We want the slope of the line. Therefore, the derivative we want is

$$\left(\frac{\partial p}{\partial T}\right)_{\Delta G.}$$

We can write this derivative in terms of quantities we know using the cyclic rule (Euler's chain relation) as,

$$\left(\frac{\partial p}{\partial T}\right)_{\Delta G} = \frac{\left(\frac{\partial \Delta G}{\partial T}\right)_p}{\left(\frac{\partial \Delta G}{\partial p}\right)_T} \tag{9}$$

We know from $dG = -SdT + Vdp$, or $d\Delta G = -\Delta SdT + \Delta Vdp$, that,

$$\left(\frac{\partial \Delta G}{\partial T}\right)_p = -\Delta S \tag{10}$$

and

$$\left(\frac{\partial \Delta G}{\partial p}\right)_T = -\Delta V \tag{11}$$

Using Equations10 and 11 in Equation 9 (the two minus signs cancel) we get,

$$\left(\frac{\partial p}{\partial T}\right)_{\Delta G} = \frac{\Delta S}{\Delta V} \tag{12}$$

Equation 12 is the Clapeyron equation. The Clapeyron equation is thermodynamically exact. It contains no approximations. There is another version of the Clapeyron equation which we obtain by inserting Equation 4 for ΔS into Equation 12,

$$\left(\frac{\partial p}{\partial T}\right)_{\Delta G} = \frac{\Delta H}{T\Delta V} \tag{13}$$

Equation 13 is useful when we want to integrate *dp* to find *p* as a function of *T*. The integration of Equation 12 or 13 is easiest if we can make the approximation that either ΔS or ΔH is reasonably constant over the temperature range. However, ΔH usually varies more slowly with temperature than does ΔS, so that the approximation is better when we integrate 13. Prepare Equation 13 for integration,

$$dp = \frac{\Delta H}{T\Delta V} dT \tag{14}$$

To integrate Equation 14 we must know how ΔH and ΔV depend on temperature. We have already said that ΔH can be regarded as approximately constant, but what about ΔV? ΔV is certainly not even near constant if one of the components of the phase transition is a gas but if the transition is solid → solid or solid → liquid we can approximate ΔV as a constant also.

Then Equation 14 integrates to,

$$p_2 - p_1 = \frac{\Delta H}{\Delta V} In \frac{T}{T_1} \tag{15}$$

If we let T_2 and p_2 range over the phase line and call them *T* and *p*, respectively, we can write pressure as a function of temperature for a solid → solid or solid → liquid phase transition line as follows,

$$p = p_1 + \frac{\Delta H}{\Delta V} In \frac{T}{T_1} \tag{16}$$

THE CLAUSIUS-CLAPEYRON EQUATION

Equations 12 and 13 are exact and are valid for all types of phase transitions, solid-liquid, solid-gas, liquid-gas, and so on. However, for solid-gas and liquid-gas phase transitions we can make two approximations which will give us a very useful equation called the Calusius-Clapeyron equation. The first approximation comes from noting that the volume of a given amount of gas is much larger that the volume of an equivalent amount of solid or liquid.

This being true, we can approximate ΔV by the volume of the gas.

That is,

$$\Delta V = V_{gas} - V_{liquid} \text{ H} - V_{gas} \tag{17}$$

or

$$\Delta V = V_{gas} - V_{solid} \text{ H} - V_{gas}. \tag{18}$$

In the second approximation we replace V_{gas} by the volume the gas would have if it were ideal. That is, we set $V_{gas} = nRT/p$. With these two approximations Equation 13 becomes,

$$\left(\frac{\partial p}{\partial T}\right)_{\Delta G} = \frac{\Delta H}{T\dfrac{nRT}{p}} = \frac{\Delta \overline{H}}{R}\frac{p}{T^2} \tag{19}$$

where the bar over the enthalpy, H, means that it is the enthalpy per mole. Equation 19 is the Clausius-Clapeyron equation. We should emphasize that Equation 19 is not exact (because it contains two approximations) and it only applies when one of the phases in the phase transition is a gas.

Equation 19 can be prepared for integration is follows,

$$\frac{dp}{p} = \frac{\Delta \overline{H}}{R}\frac{dT}{T^2}. \tag{20}$$

Equation 20 can be integrated if we know how ΔH changes with temperature or, as we will now do, if we regard ΔH as approximately constant. With the ΔH regarded as constant Equation 20 integrates to,

$$In\frac{p_2}{p_1} = \frac{\Delta \overline{H}}{R}\left(\frac{1}{T^2} - \frac{1}{T_1}\right) \tag{21}$$

Equation 21 is the integrated form of the Clausius-Clapeyron equation. If we want pressure as a function of temperature we can let the point (p_2, T_2) range over the phase line and call it simply (p, T) and then take the antilog of both sides,

$$p = p_1 e^{-\frac{\Delta H}{R}\left(\frac{1}{T}-\frac{1}{T_1}\right)} \qquad (22)$$

Notice that if we know two points on the curve we can solve for ΔH, or if we know one point on the curve and ΔH we can solve for any other point or even get the entire curve. Equation 22 is very useful for estimating vapour pressures at various temperatures when ΔH and the vapour pressure at one temperature.

You can get fancy with Equations 19 and 20 if you want. For example, we know that ΔH for a phase transition is not really constant. If we wanted to take into account the temperature dependence of ΔH we need to know the heat capacities of the two phases, or rather the difference in the heat capacity for the two phases,

$$\Delta C_p = C_{p\beta} - C_{p\alpha} \qquad (23)$$

Then

$$\Delta\overline{H}(T) = \Delta\overline{H}(T_1) + \int_{T_1}^{T} \Delta\overline{C}_p dT' \qquad (24)$$

Initially we regarded ΔH as constant. The next level of approximation would be to regard ΔC_p as constant so that,

$$\Delta\overline{H}(T) = \Delta\overline{H}(T_1) + \Delta\overline{C}_p(T - T_1) = \Delta\overline{H}(T_1) - \Delta\overline{C}_p T_1 + \Delta\overline{C}_p T \quad (25)$$

Using this expression for ΔH in Equation 20 we get

$$\frac{dp}{p} = \frac{\Delta\overline{H}(T_1) - T_1\Delta\overline{C}_p dT}{R}\frac{dT}{T^2}, \qquad (26)$$

$$\frac{dp}{p} = \frac{\Delta\overline{H}(T_1) - T_1\Delta\overline{C}_p}{R}\frac{dT}{T^2} + \frac{\Delta\overline{C}_p}{R}\frac{dT}{T} \qquad (27)$$

Equation 27 integrates to,

$$In\frac{p}{p_1} = \frac{\Delta\overline{H}(T_1) - T_1\Delta\overline{C}_p}{R}\left(\frac{1}{T} - \frac{1}{T_1}\right) + \frac{\Delta\overline{C}_p}{R} In\frac{T}{T_1} \qquad (28)$$

You can get even fancier. The next level of approximation would be to include the temperature dependence of the heat capacities, but we won't deal with that here.

Equation 12 does not depend on ΔG being zero. It only assumes that ΔG is constant for the α -β transition. Of course, if ΔG is not zero then equation 4 is not true which makes

Equation 13 invalid. Further, if ΔG is not zero then either the α phase or the β phase is thermodynamically unstable. (We usually use the word *metastable* to describe this situation. Examples are supercooled liquids, superheated liquids, supercooled vapours, and some metastable solids - like diamond.) Nevertheless, one can visualize a family of curves running along side the phase diagram curve with ΔG constant at some value other than zero.

MELTING CURVE FOR WATER; VAPOUR PRESSURE

MELTING CURVE FOR WATER

We mentioned in our initial discussion of one-component phase diagrams that the slope of the melting curve for water was negative.

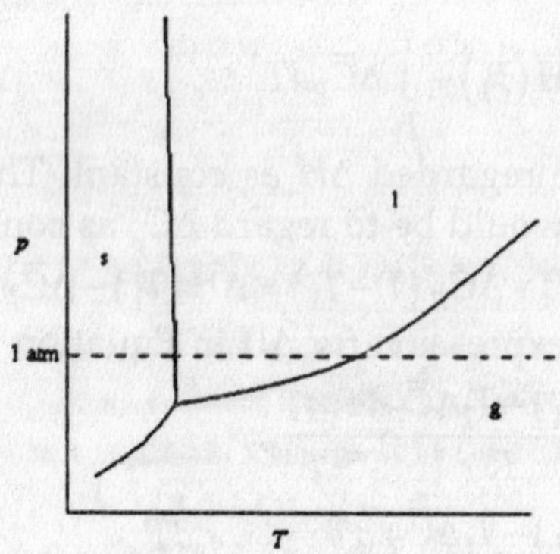

Water is one of just a few known substances with this feature (bismuth is another), the melting curve for most substances has a positive slope. Let us consider the thermodynamics consequences of this negative slope. Since we are talking about the melting curve we must use the Clapeyron equation,

$$\left(\frac{\partial P}{\partial T}\right)_{\Delta G} = \frac{\Delta S}{\Delta V}, \qquad (1)$$

where, of course, the ΔS and ΔV refer to the melting process or the freezing process. For the purposes of this discussion let us consider the melting process.

That is, we will assume that the phase change is, $H_2O(s)$-$H_2O(l)$.

A negative slope on the phase diagram means that the derivative in Equation 1 is negative.

This in turn requires that,

$$\left(\frac{\partial P}{\partial T}\right)_{\Delta G} = \frac{\Delta S}{\Delta V} < 0. \tag{2}$$

We know that ΔS for the melting process is positive because the system is going from the highly ordered crystalline solid state to the more disordered liquid state (and we can calculate it from $\Delta S = \Delta H_{fusion}/T_{mp}$). Therefore, the only way the derivative in Equation 2 can be negative is if ΔV is negative. This means that the molar volume of the liquid is *smaller* than the molar volume of the solid at the melting point,

$$\overline{V}_l - \overline{V}_S < 0,$$
$$\overline{V}_l < \overline{V}_S \tag{3a, b}$$

Taking the reciprocal of Equation 3b we find that,

$$\frac{1}{\overline{V}_l} > \frac{1}{\overline{V}_S} \tag{4}$$

and, since the density is proportional to the reciprocal of the volume, this means that the density of the liquid is greater than the density of the solid (at the melting point).

This result is counterintuitive. We expect that materials will expand on melting and most materials do expand on melting. However, water is unusual in this respect and contracts on melting. The visible consequence of this fact is that ice floats on liquid water. We are so accustomed to this that we do not even notice that it is unusual. However, if you have worked in the laboratory with other solid/liquid systems you will be aware that most often the solid sinks in the melt.

You can argue that the slope of the melting curve of water makes a crucial contribution to the ability of our planet, Earth, to support life. If the slope were positive then ice would sink in water. In cold climates ice would form in winter and sink to the bottom. This ice would be insulated from the summer warmth by the layer of water above it and would only partially melt. The next winter more ice would accumulate on the bottom. Eventually, all rivers and lakes and possibly a large portion of the oceans would be frozen from the bottom up with only a shallow layer of liquid water being formed on the surface during the summer. There would likely be no animal or plant life in these frozen lakes and rivers.

VAPOUR PRESSURE

We mentioned in our discussion of one-component phase diagrams that the vapourization curve is sometimes referred to as the vapour pressure curve. This is because this curve - along with the sublimation curve - gives a plot of the vapour pressure of the substance versus temperature.

The vapour pressure is a measure of the ability of molecules to escape from the surface of a solid or liquid. The following discussion uses ideas that are outside the scope of thermodynamics, but they may be useful in calibrating our intuition. A solid or liquid is held together by intermolecular attractive forces. However, the molecules have kinetic energy (energy of motion) which we will later see is proportional, on the average, to the Kelvin temperature. There is a competition. The attractive forces are trying to keep the molecules in the solid or liquid and the kinetic energy is trying to take them out. The vapour pressure is the net result of this competition.

As more and more molecules build up in the space above the solid or liquid more of them will collide back with the surface and some of them will stick. At a fixed temperature the system will reach a state of "dynamic equilibrium," where just as many molecules escape from the surface as collide back and stick. At this point the pressure exerted by the molecules in the space above the solid or liquid is, by definition, the vapour pressure of the solid or liquid.

As the temperature is increased the average kinetic energy increases so more molecules are able to overcome the intermolecular forces and the vapour pressure increases.

INCREASING THE VAPOUR PRESSURE BY THE APPLICATION OF AN EXTERNAL PRESSURE

One can also increase the vapour pressure of a solid or liquid by applying an external pressure. This effect is not as easy to understand qualitatively as the effect of temperature, but it is easy to calculate thermodynamically. Recall that for solids and liquids, which are not very compressible over relatively small pressure ranges, the Gibbs free energy change is given by,

$$\Delta G \approx V\left(p-p_o\right). \tag{5}$$

In words, Equation 5 says that G increases when we "squeeze" a solid or a liquid. (The Gibbs free energy of a gas also increases when you "squeeze it, but gases are quite compressible so that Equation 5 does not hold for gases.) We would expect that if the Gibbs free energy of a solid or liquid increases its vapour pressure would also increase. We shall see if this is true. Rewrite Equation 5 as follows,

$$\Delta G = G_p - G_{p_o} = V\left(p-p_o\right). \tag{6}$$

We will rearrange Equation 6 and then divide it by the number of moles of substance, n.

$$G_p = G_{po} + V\left(p-p_o\right),$$

$$(G_p = G_{po} + V\left(p-p_o\right))\frac{1}{n},$$

$$\frac{G_p}{n} = \frac{Gp_o}{n} + \frac{V}{n}\left(p-p_o\right),$$

$$\mu_p = \mu_{p_o} + \overline{V}\left(p-p_o\right). \tag{7a, b, c, d}$$

In Equation 7d we have used the fact that the Gibbs free energy per mole is called the chemical potential and is symbolized by a lower case Greek "mu."

Let us consider two systems, all at temperature, T. The first system is a closed container which contains only, say, liquid water and water vapour in equilibrium. Since the system is in equilibrium the chemical potential of the liquid water and water vapour must be equal. Using the approximation that the vapour is an ideal gas we can write,

$$\mu_l^o = \mu_g,$$

$$\mu_l^o = \mu_g^o, +RT\,In\frac{p_o}{1atm}. \tag{8a, b}$$

(In Equation 8b we have used the expression for the

variation of the chemical potential of an ideal gas with pressure.)

Our second system consists of a similar container of liquid water and water vapour, but also contains an inert gas, say air, at a pressure, P. It is the air that is squeezing the liquid and providing the "external pressure." We will call this system the "prime" system. In this second system the chemical potentials of the liquid water and water vapour must still be equal, but they are not equal to the same values as in the first system. In the "prime" system we have,

$$\mu'_1 = \mu'_g, \tag{9}$$

but

$$\mu'_i = \mu^o_i + \overline{V}(P-0) = \mu^\circ + \overline{V}P \tag{10}$$

and

$$\mu'_g = \mu^o_g + RT\,In\frac{p'}{1\,atm}. \tag{11}$$

So we have two equations equating chemical potentials in the gas and liquid phases, one for the first system and one for the second system.

$$\mu^\circ_i + \overline{V}P = \mu^\circ_g + RT\,In\frac{p'}{1\,atm}, \tag{12}$$

and

$$\mu^\circ_i = \mu^\circ_g + +RT\,In\frac{p_\circ}{1\,atm}. \tag{12b}$$

When we subtract Equation 8b from Equation 12 we eliminate all the standard state chemical potentials and are left with

$$\overline{V}P = RT\,In\frac{p'}{1\,atm} - RT\,In\frac{p_\circ}{1\,atm}$$

$$= RT\,In\frac{p'}{p_\circ}, \tag{13a, b}$$

so that

$$p' = p_\circ e^{\frac{\overline{V}p}{RT}}. \tag{14}$$

Equation 14 tells us that the vapour pressure of the liquid, in this case water, increases when we apply an external pressure. For water the vapour pressure increases by 0.074 per cent at 1.00 atm and 7.6 per cent at 100 atm.

MIXTURES; PARTIAL MOLAR QUANTITIES; IDEAL SOLUTIONS

MIXTURES

We have seen that the combined first- and second laws for an open system is written,

$$dU = TdS - pdV + \mu_1 dn_1 + \mu_2 dn_2 + \ldots, \tag{1}$$

Where the initial definitions of the μ_i were given by,

$$\mu_1 = \left(\frac{\partial U}{\partial n_1}\right)_{S,V,n_2,\ldots} \tag{2}$$

An equivalent, and more easily understood, definition of the μ_i was given by,

$$\mu_1 = \left(\frac{\partial G}{\partial n_1}\right)_{T,p,n_2,\ldots} \tag{3}$$

μ_i is called the "chemical potential" and it can be described in words as the Gibbs free energy per mole of substance. Another name for this quantity is "partial molar Gibbs free energy." We found, using the integrated form of Equation 1 and the definition of the Gibbs free energy, that the Gibbs free energy could be written as,

$$G = \mu_1 n_1 + \mu_2 n + \ldots \tag{4}$$

Equation 4 is an exact thermodynamic equation and contains no approximations.

Notice in Equation 3 that μ_i is intensive, it is given by the extensive G divided by the extensive n_i. Also, the variables that are being held constant are temperature and pressure along with the number of moles of the other components of the system. We can define an equation similar to Equation 3 for any extensive property. The partial molar Gibbs free energy is the most important of these because it provides a measure of the "driving force" for chemical processes. The next most important of these quantities is the partial molar volume,

$$\overline{V}_i = \left(\frac{\partial V}{\partial n_i}\right)_{T,p,n,...} \quad (5)$$

(We will use the bar over the symbol to indicate partial molar quantities. Some texts use a subscript, m, as in V_{mi} to indicate them.) The partial molar volume in Equation 5 can be thought of in several ways. It is the incremental volume obtained by adding a small amount of component *i* to the mixture while holding the temperature, pressure, and the number of moles of all the other components constant divided by the number of moles of component *i*. Another way to look at it is to say that it is the incremental volume obtained by adding one mole of component *i* to an infinite sample of the mixture. The partial molar volume is not necessarily the same as the volume of one mole of the pure component. (The intermolecular interaction between molecules of one component and molecules of other components may be different than the interaction of molecules of a component with other molecules of the same component.)

Let us regard *V* as a function of temperature, pressure and composition:

$$V = V(T, p, n_1 n_2 ...).$$

Then

$$dV = \left(\frac{\partial V}{\partial T}\right)_{p,n} dT + \left(\frac{\partial T}{\partial n}\right)_{T,n} dp + \left(\frac{\partial V}{\partial n_1}\right)_{T,p,n_2..} dn_1 + \left(\frac{\partial V}{\partial n_2}\right)_{T,p,n_1} dn_2 + ...$$

$$\left(\frac{\partial V}{\partial T}\right)_{p,n} dT + \left(\frac{\partial T}{\partial n}\right)_{T,n} dp + \overline{V_1} dn_1 + \overline{V_2} dn_2 + ... \quad (6a, b)$$

If we hold *T* and *p* constant we get

$$dp + \overline{V_1} dn_1 + \overline{V_2} dn_2 + ..., \quad (7)$$

which can be integrated (similar to the way we integrated Equation 1, above) to give,

$$V = n_1 \overline{V}_1 + n_2 \overline{V}_2 + ...,$$
$$= \sum_i n_i \overline{V}_i \quad (8a, b)$$

The volumes given in Equations 7, 8a, and 8b are the partial molar volumes which we have already said are not necessarily equal to the molar volumes of the pure components. Thus, Equations 8a and b tell us that volumes may not be additive. That is, if we were to mix one liter of pure ethanol with one liter of pure water the final volume of the mixture would not likely be two liters. That is because water molecules interact with ethanol molecules differently than they interact with other water molecules.

As we have said above, we can define a partial molar quantity from any extensive variable. Thus we can define the partial molar entropy as,

$$\overline{S}_i = \left(\frac{\partial S}{\partial n_i}\right)_{T,p,n_j}, \tag{9}$$

or the partial molar enthalpy as,

$$\overline{H}_i = \left(\frac{\partial H}{\partial n_i}\right)_{T,p,n_j}, \tag{10}$$

It is also true that

$$S = n_1\overline{S}_1 + n_2\overline{S}_2 + \ldots, \tag{11}$$

and so on.

How to Measure Partial Molar Volumes

There are several ways that partial molar volumes can be measured. One way is to begin with one mole of a compound, call it component 1, add a small amount of component 2 and measure the volume, add a little more of component 2 and measure the volume again. Keep doing this until the desired concentration range has been covered.

Then fit the volume data to a curve, for example, of the form,

$$V\left(n_1 = 1, n_2\right) = \overline{V_1^\circ} + an_2 + bn_2^{3/2} + cn_2^2 + \ldots \tag{12}$$

(Why would we not want to include that 1/2 power of n_2 in Equation 12? People use whatever powers of n_2 they need fit the data.) The constants, *a, b, c,* etc are obtained from the curve fitting and the first term is the molar volume of pure component 1. Then the partial molar volume of component 2 can be obtained by direct differentiation,

$$\left(\frac{\partial V}{\partial n_2}\right)_{T,p,n_1} = \overline{V_2}$$

$$= 0 + a + \frac{3}{2}bn_2^{1/2} + 2cn_2 + ... \qquad (13a, b)$$

The partial molar volume of component 1 can be obtained from,

$$V = n_1\overline{V}_1 + n_2\overline{V}_2$$

$$= \overline{V}_1 + n_2\overline{V}_2, \qquad (14a, b)$$

or

$$\overline{V_1} = \frac{V - n_2\overline{V}_2}{n_2}$$

$$= V - n_2\overline{V}_2. \qquad (15a, b)$$

(The last step is because, in this case we have set $n_1 = 1$.)

IDEAL SOLUTIONS

We will define an ideal solution as a solution for which the chemical potential of each component is given by,

$$\mu_i = \mu_i^* + RT\, In\, X_i, \qquad (16)$$

where μ_i^* is the chemical potential of pure component *i*, and X_i is the mole fraction of component *i* in the solution.

(Many texts define an ideal solution as a solution which obeys Raoult's law over the full range of composition,

$$p_i = X_i p_i^*, \qquad (17)$$

where p_i^* is the vapour pressure of pure component *i*.)

We will now prove that an ideal solution obeys Raoult's law (using our definition of an ideal solution).

Consider a solution of two components where the mole fraction of component 1 is X_1. We know that the chemical potential of component 1 must be the same in the solution as in the vapour in equilibrium with the solution. That is,

$$\mu_{1l} = \mu_{1g}, \qquad (18)$$

but the solution is ideal so,

$$\mu_{1l} = \mu_{1/l}^* + RT\, In\, X_1. \qquad (18a)$$

Also, we can approximate the vapour as an ideal gas so,

$$\mu_{1g} = \mu_{1g}^{\circ} + RT\, In \frac{p_1}{1\,atm}, \tag{19}$$

where p_1 is the vapour pressure (partial pressure) of component 1 above the solution. Combining Equations 16, 18, and 19 we get,

$$\mu_{1/l}^{*} + RT\, In\, X_1 = \mu_{1g}^{\circ} + RT\, In \frac{p_1}{1\,atm}. \tag{20}$$

Equation 20 doesn't help us very much all by itself. However we have some more information. We know that for the pure component 1 we have $X_1 = 1$, and we know that the pressure of component 1 vapour in equilibrium with the liquid is just the vapour pressure of the pure liquid, p_1^*, so that,

$$\mu_{1/l}^{*} = \mu_{1g}^{\circ} + RT\, In \frac{p_1^{*}}{1\,atm}. \tag{21}$$

Let us now subtract Equation 21 from Equation 20 to get

$$RT\, In\, X_1 = RT\, In \frac{p_1}{1\,atm} - RT\, In \frac{p_1^{*}}{1\,atm}, \tag{22}$$

from which we conclude that,

$$In\, X_1 = In \frac{p_1}{p^{*}},$$

$$X_1 = \frac{p_1}{p^{*}}, \tag{23a, b}$$

or

$$p_1 = X_1 p^{*}, \tag{17}$$

which is Raoult's law.

CALCULATION USING RAOULT'S LAW

Benzene and toluene form a solution which is very nearly ideal. Consider a mixture of benzene (Bz) and toluene (Tol) at 60° C. At 60° C the vapour pressures of pure benzene and pure toluene are 385 Torr and 139 Torr, respectively. What are the vapour pressures of benzene and toluene in a mixture with $X_{Bz} = 0.400$, and $X_{Tol} = 0.600$, and what is the composition of the vapour in equilibrium with this solution?

Use Raoult's law to find the vapour pressures of the two species,

$$P_{Bz} = X_{Bz}P^*_{Bz} = 0.400 \times 385\,Torr = 154\,Torr,$$

$$P_{Total} = X_{Tol}P^*_{Tol} = 0.600 \times 139\,Torr = 83.3\,Torr.$$

The total pressure is the sum of these two individual pressures, 237 Torr.

The composition of the vapour phase is obtained from the vapour pressures and Dalton's law of partial pressures,

$$X_{Bz} = \frac{P_{Bz}}{P_{Tot}} = \frac{154}{237} = 0.649,$$

$$X_{Tol} = \frac{P_{Tol}}{P_{Tol}} = \frac{83.3}{237} = 0.351.$$

Notice that the composition of the vapour is not the same as the composition of the liquid, the vapour phase is much richer in the more volatile compound, benzene. This fact will be important when we discuss vapour pressure diagrams and two-component phase diagrams.

PROPERTIES OF IDEAL SOLUTIONS

Given the chemical potentials for the components of an ideal solution we can calculate a number of properties of ideal solutions. For example, the Gibbs free energy of mixing is the easiest to calculate,

$$\Delta G_{mix} = G_{mixed} - G_{unmixed},$$

$$n_1\mu_1 + n_2\mu_2 + \ldots - n_1\mu^*_1 + n_2\mu^*_2 - \ldots.$$

$$n_1\mu^*_1 + n_2\mu^*_2 - \ldots$$

$$RT\left(n_1\, In X_1 + n_2\, In\, X_2 + \ldots\right). \qquad (24a, b, c, d)$$

To bring this equation into the usual form multiply and divide by the total number of moles, n, and bring the $1/n$ inside the parentheses to convert the number of moles of each component into a mole fraction,

$$\Delta G_{mix = nRT(X_1 In X_1 + X_2 In X_2 + \ldots)}. \qquad (25)$$

Notice that the Gibbs free energy of mixing is negative, as one would expect for a spontaneous process at constant temperature and pressure.

We can also calculate other properties of mixing ideal solutions.

The entropy of mixing is,

$$\Delta S_{mix} = -\left(\frac{\partial \Delta G_{mix}}{\partial T}\right)_{p_i n_j}$$

$$= -nR\left(X_1 \, In X_1 + X_2 \, In \, X_2 + ...\right), \tag{26}$$

which is the same as the entropy of mixing for ideal gases.

The volume change on mixing can be found from,

$$\Delta V_{mix} = \left(\frac{\partial \Delta G_{mix}}{\partial p}\right) T, n_i = 0, \tag{27}$$

so that volumes are additive for an ideal solution. That is, if we were to mix one liter of benzene and one liter of toluene the final volume of the solution would be two liters.

We can also determine whether or not there is any heat of reaction.

$$\Delta G_{mix} = \Delta G_{mix} + T\Delta S_{mix}$$

$$= nRT\left(X_1 In X_1 + X_2 In X_2 + ...\right)$$

$$= +T(-nR)(X_1 In X_1 + X_2 In X_2)$$

$$= 0. \quad (28)$$

That is, there is no heat involved in the mixing of ideal solutions. If we mix several components and the mixture gets hot or cold we can be sure that the solution formed was not ideal.

ACTIVITY AND ACTIVITY COEFFICIENTS

We have now seen and used several expressions for the chemical potential of a substance or component in a mixture.

For a one-component ideal gas we had,

$$\mu = \mu^\circ + RT In \frac{p}{p^\circ}, \tag{1}$$

where μ^o is the chemical potential when $p = p^o$ and p^o is usually one atmosphere.

For a mixture of ideal gases it can be shown that for each component, *i*, the chemical potential is given by,

$$\mu_i = \mu_i{}^\circ + RTIn\frac{p_i}{p^\circ}, \tag{2}$$

For a nonideal gas we used the fugacity, f, instead of the pressure and the chemical potential for one component is,

$$\mu = \mu^\circ + RTIn\frac{f(p)}{p^\circ}. \tag{3}$$

For a mixture of nonideal gases it can be shown that,

$$\mu_i = \mu_i{}^\circ + RTIn\frac{f_i}{p^\circ}, \tag{4}$$

only now the fugacity of component i is a function of the pressures of all the gases in the mixture,

$$f_i = f_i(p_1, p_2, p_3, ...). \tag{5}$$

For ideal solutions we found that,

$$\mu_i = \mu_i{}^\circ + RTInX_i. \tag{6}$$

These expressions for chemical potential all have the form of a reference or standard state chemical potential plus RT times the logarithm of something related to pressure or concentration.

This form turns out to be very important, so important that G. N. Lewis used it to give the most general case of chemical potential as,

$$\mu_i = \mu_i{}^\circ + RTIna_i. \tag{7}$$

The quantity, a_i, is called the "activity" of component i and Equation 7 should be regarded as the definition of activity. Notice that the activity has no units.

All of the special cases we have been considering so far can be reconciled to this definition of activity. Thus, for an ideal gas mixture,

$$a_i = \frac{p_i}{p^\circ};$$

for a nonideal gas mixture,

$$a_i = \frac{f_i}{p^\circ};$$

for an ideal solution,

$$a_i = X_i,$$

and so on. In a nonideal solution we would have to just write Equation 7 again,

$$\mu_i = \mu_i^\circ + RT\ln a_i. \tag{7}$$

In Equation 7 all the nonidealities of the solution are absorbed into the activity. We will see a more convenient way to write this below under the heading "activity coefficient."

We can find the activity of a component of a nonideal solution from measurements of the vapour pressure of that component in the vapour in equilibrium with the solution.

We know that the chemical potential of a component must be the same in the vapour as in the liquid. that is, from Equations 2 and 7 we obtain,

$$\mu_{il} = \mu_{ig} \tag{8a, b}$$

$$\mu^*_{il} = RT\ln a_i = \mu^\circ_{ig} + RT\ln \frac{p}{p^\circ},$$

but for pure component *i*, we must have,

$$\mu^*_{il} = \mu^\circ_{ig} + RT\ln \frac{p_i^*}{p^\circ}. \tag{9}$$

Note: When dealing with liquid solutions it is customary to write the chemical potential of the pure liquid as μ^*_{il} instead of the usual μ^o_{il}, which means that the standard state for a liquid is the pure liquid itself. p_i^* is the vapour pressure of the pure liquid.

Subtracting Equation 9 from Equation 8b yields,

$$RT\ln a_i = RT\ln \frac{p_i}{p^\circ} - RT\ln \frac{p_i^*}{p^\circ}$$

$$RT\ln \frac{p_i}{p_i^*}, \tag{10a, b}$$

so that,

$$a_i = \frac{p_i}{p_i^*}. \tag{11}$$

If the solution were ideal p_i would be given by Raoult's law and the activity would be just the mole fraction.

ACTIVITY COEFFICIENT

Sometimes it is convenient to write the activity as the product of an ideal part times a nonideality correction part.

For example, in a real solution we might write,

$$a_i = X_i \gamma_i, \tag{12}$$

or, in a real gas we would write,

$$f_i = p_i \gamma_i, \tag{13}$$

so that,

$$a_i = \frac{f_i}{p^\circ} = \frac{p_i \gamma_i}{p^\circ} = \frac{p_i}{p^\circ}\gamma_i. \tag{14}$$

In the latter case we already know that

$$f(p) = pe^{\int_0^p \frac{(Z-1)}{p'}dp'}, \tag{15}$$

so that

$$\gamma = e^{\int_0^p \frac{(Z-1)}{p'}dp'}. \tag{16}$$

In the case of the nonideal gas $\gamma \to 1$ as $p \to 0$. In the case of a nonideal solution $\gamma_i \to 1$ as $X_i \to 1$. From Equations 11 and 12 we see that,

$$a_i = \frac{p_i}{p_i^*} = X_i \gamma_i, \tag{17}$$

from which we conclude that for a nonideal solution,

$$\gamma_i = \frac{p_i}{X_i p_i^*} = \frac{\textit{actual vapor pressure}}{\textit{Raoult's law vapor pressure}}. \tag{18}$$

In general, the activity coefficient is a unitless parameter that contains all of the nonideality of a system.

VAPOUR PRESSURE AND BOILING POINT

We are now ready to begin talking about phase diagrams involving two components. Our first few phase diagrams will involve only the liquid and gas (or vapour) phases. Later we will discuss two-component phase diagrams involving liquids and solids.

Consider a solution of two liquids, say benzene and toluene held at a constant temperature. Benzene and toluene form very nearly ideal solutions so we shall regard the solution as ideal. We know that an ideal solution obeys Raoult's law, so for each component we must have,

$$p_i = X_i p_i^*, \qquad (1)$$

where p_i is the partial pressure of component i above the liquid mixture (or the vapour pressure of i in the mixture), X_i is the mole fraction of component i in the liquid, and p_i^* is the vapour pressure of pure component i.

If we now hold temperature constant and plot the vapour pressure of a benzene-toluene solution as a function of composition we would get a graph that looked something like this,

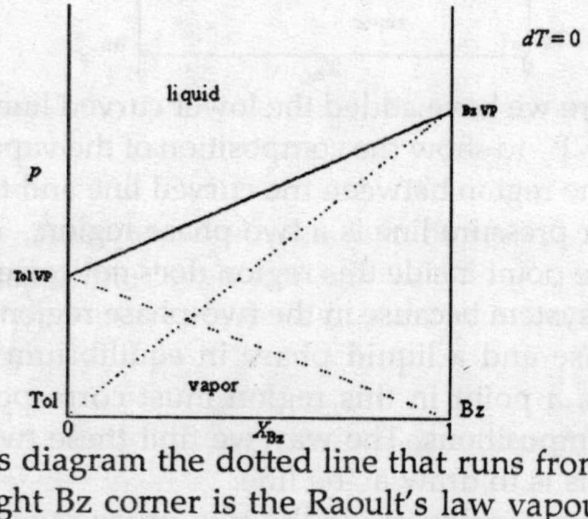

On this diagram the dotted line that runs from "Tol VP" to lower right Bz corner is the Raoult's law vapour pressure of toluene. The dotted line that runs from the lower left Tol corner to "Bz VP" is the Raoult's law vapour pressure of benzene. The solid line from "Tol VP" to "Bz VP" is the total vapour pressure of the solution which is just the sum of the two Raoult's law vapour pressures (the sum of two straight lines is a straight line).

There is no indication on the above diagram what the composition of the vapour is at each pressure. We know that at any given pressure the vapour will be richer in the more volatile component, in this case richer in benzene. In our initial discussion of Raoult's law we worked a problem in which the composition of a benzene-toluene solution was 0.400 mole fraction benzene, but the vapour in equilibrium with the liquid was 0.649 mole fraction benzene. This means that we must draw another line to indicate the composition of the vapour at each pressure.

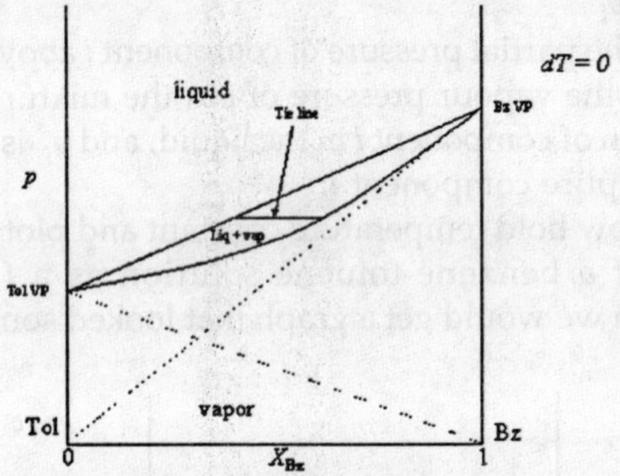

In figure we have added the lower curved line from "Tol VP" to "Bz VP" to show the composition of the vapour at each pressure. The region between the curved line and the straight total vapour pressure line is a two phase region.

A single point inside this region does not correspond to a state of the system because in the two-phase region we have a vapour phase and a liquid phase in equilibrium with each other. Thus, a point in this region must correspond to two different compositions. The way we find these two different compositions is to draw a "tie line."

Although a point inside the two phase region does not give us the composition of either the liquid or the vapour phase, it does give the overall composition of the whole system. The tie line is a horizontal line through that point which intersects the two boundary lines of the two phase region.

The intersection of our horizontal line (the tie line) with the boundary line on the liquid side of the region tells us the composition of the liquid and the intersection of the tie line with the boundary curve on the vapour side tells us the composition of the vapour phase.

Now we have to deal with nonideal solutions. We will

distinguish two cases: positive deviations from Raoult's law and negative deviations from Raoult's law. Nonideal solutions which have positive deviations from Raoult's law have vapour pressures which are higher than Raoult's law would predict.

Nonideal solutions which have negative deviations from Raoult's law have vapour pressures which are lower than Raoult's law would predict.

This is because the attractive forces between the unlike molecules are stronger than the attractive forces between like molecules. A vapour pressure diagram for a system with positive deviations from Raoult's law might look qualitatively like the following:

Positive deviations from Raoult's law

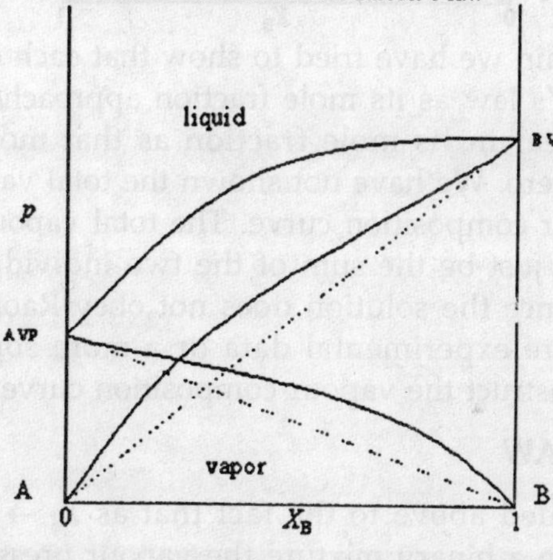

This vapour pressure diagram shows qualitatively how the vapour pressures vary with composition for positive deviations from Raoult's law.

We have not tried to draw the vapour composition curve on the diagram. We have tried to show that each component obeys Raoult's law in the limit as its mole fraction goes to one. Also, the vapour pressures of the individual components becomes linear in mole fraction as that mole fraction approaches zero. This latter phenomenon is known as "Henry's law" and will discussed further below. Negative deviations from Raoult's law might look qualitatively like the following diagram.

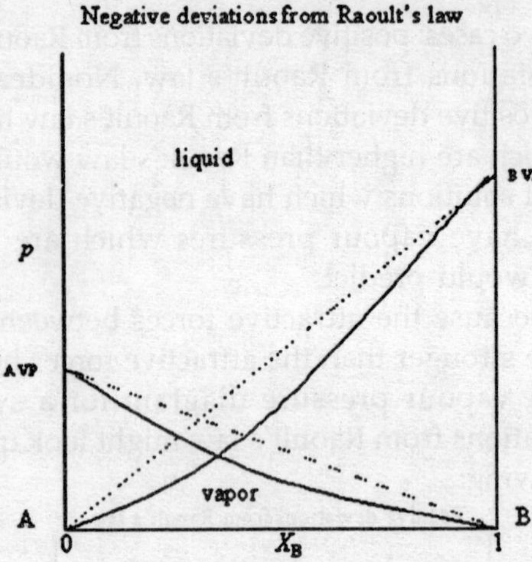

Once again we have tried to show that each component obeys Raoult's law as its mole fraction approaches one and becomes linear in its mole fraction as that mole fraction approaches zero. We have not shown the total vapour curve or the vapour composition curve. The total vapour pressure curve would just be the sum of the two individual vapour pressures. Since the solution does not obey Raoult's law it would require experimental data or a more sophisticated theory to construct the vapour composition curve.

HENRY'S LAW

We alluded above to the fact that as $X_i \to 1$ for each component in a binary mixture the vapour pressure of that component obeys Raoult's law.

That is,

$$p_i = X_i p_i^* \tag{1}$$

even for nonideal solutions is the limit as $X_i \to 1$.

However, at the other end of the scale, for small mole fractions, p_i becomes proportional to X_i, but the proportionality constant is not p_i^*.

That is, for small mole fractions

$$p_i = K_i X_i^* \tag{2}$$

Equation 2 is called Henry's law and the constant, K_i, is the Henry's law constant. The Henry's law constant depends on the nature of both the solvent and the solute.

It can be shown thermodynamically that in a concentration region where one component obeys Raoult's law the other will necessarily obey Henry's law.

Henry's law is most often used to describe the solubility of gases in liquids.

Example Calculation

Let's calculate the solubility of N_2 gas in water at 25° C and 4.0 atm pressure. From a table of Henry's law constants we find the constant for nitrogen gas in water is 6.51×10^7 Torr.

From equation 2 we get,

$$X_{N_2} = \frac{P}{K}$$

$$= \frac{4.0 \times 760}{6.51 \times 10^7}$$

$$= 4.7 \times 10^{-5} \quad \text{(3a, b, c)}$$

This amounts to about 0.0026 mol of N_2 per liter of water or 0.073 g/L.

One can use Henry's law to calculate the amount of oxygen gas dissolved in water which is exposed to the atmosphere. The Henry's law constant for O_2 gas in water at 25°C is 3.3×10^7. Since air is about 21 mole per cent oxygen, at an air pressure of one atmosphere the partial pressure of O_2 would be about 0.21 atm. It is easy to use Henry's law to calculate the oxygen content of water in contact with air at one atmosphere pressure,

$$X_{O_i} = \frac{P}{K}$$

$$= \frac{0.21 \times 760}{3.30 \times 10^7}$$

$$= 4.8 \times 10^{-6} \quad \text{(4a, b, c)}$$

This corresponds to about 0.00027 moles of O_2 per liter of water, or 0.0086 g/L This does not seem like very much oxygen dissolved in water, but we note two things: Thermodynamically, the chemical potential of the dissolved O_2 is the same as the chemical potential of the O_2 in the gas in equilibrium with the water solution. Second, this seems to be enough to keep fish alive because fish use their gills to extract the oxygen and use it to metabolize food and produce energy.

There are other issues involved in keeping fish alive in water. At equilibrium there is enough oxygen to keep fish alive, but as the fish remove the oxygen with their gills the concentration of oxygen decreases. The dissolution and diffusion of O_2 through water are relatively slow processes. Unless a way is found to replenish the oxygen the fish will die. In large bodies of water this is no problem, but in small fish tanks it can be fatal to the fish. This is why most fish tanks have a "bubbler" in them. The bubbler increases the exposure of the water to the oxygen in air and gently stirs the water to hasten the saturation of water with oxygen.

One other comment about the solubility of gases in water. A condition known to deep sea divers as the "bends" is caused by the solubility of nitrogen in water. We know from Henry's law that the solubility increases with pressure. At the high pressures produced by deep sea diving [and some scuba diving] nitrogen gas dissolves in the blood. If the diver rises from the depths very slowly, over a period of many hours, the blood will gradually release the nitrogen and remain at equilibrium with the surrounding gas pressure. However, if the diver rises too fast the nitrogen will come out of solution as the pressure is released and form bubbles of nitrogen gas in the blood and other body fluids. The bubbles accumulate at the joints, producing great pain, and causing the diver to "double over in pain," Thus the condition is referred to as "the bends." The bends can even lead to death unless a way is found to redissolve the nitrogen gas. Commercial and military diving support ships carry a decompression chamber so that a diver can be brought up quickly and then recompressed in the chamber.In the decompression chamber the pressure can be released gradually allowing the nitrogen to escape the blood stream without forming the bubbles and making it unnecessary to remain at the diving site to bring the diver up slowly.

BOILING DIAGRAMS

The diagrams shown above, where we are plotting vapour pressures as a function of composition at constant temperature are not the must useful diagrams. Since we live in a more-or-less constant pressure environment it is more useful and interesting to plot the properties of the solution as temperature versus composition.

Think of the vapour pressure diagrams we showed above as slices at constant temperature through a three-dimensional diagram showing the properties of the mixture as a function of temperature, pressure, and composition.

The lines in the above vapour pressure diagrams will appear on this three-dimensional diagram as surfaces, and the points, such as boiling points will be curved lines. (Because the boiling point changes with changing pressure.)

Now, instead of taking slices at constant temperature, let's take slices through our three-dimensional diagram at constant pressure. If we take our slice at a pressure of one atmosphere the new diagram, called a boiling diagram, would look something like,

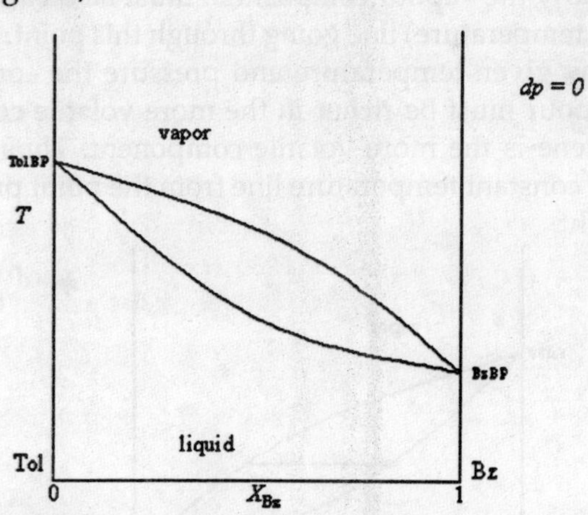

This diagram is called a "boiling diagram." Notice that at the edges we have points representing the boiling points of the pure components. The lower curve gives the boiling point of the liquid mixture as a function of composition.

Suppose we pick a point on the lower curve at, say, a mole fraction of benzene of about 0.33.

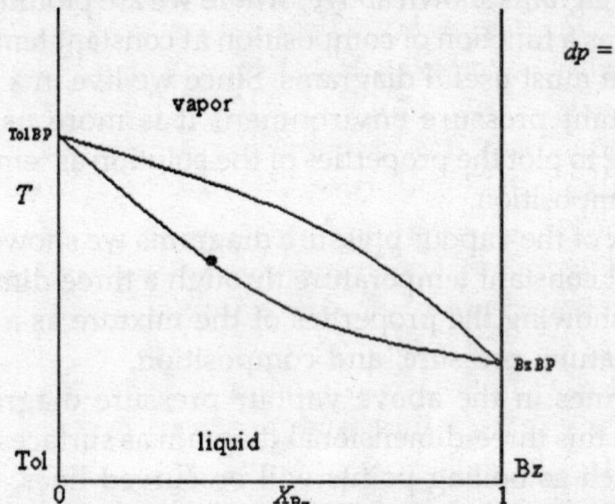

This point gives the boiling point of a mixture which is 0.33 mole per cent benzene. We can ask, what is the composition of the vapour which is in equilibrium with the liquid at this temperature? Since we are asking a question about something at this temperature, the temperature of the point, clearly the vapour composition must lie on a horizontal (constant temperature) line going through this point. We know that at any given temperature and pressure the composition of the vapour must be richer in the more volatile component and benzene is the more volatile component. Thus we must draw our constant temperature line from the point proceeding to the right.

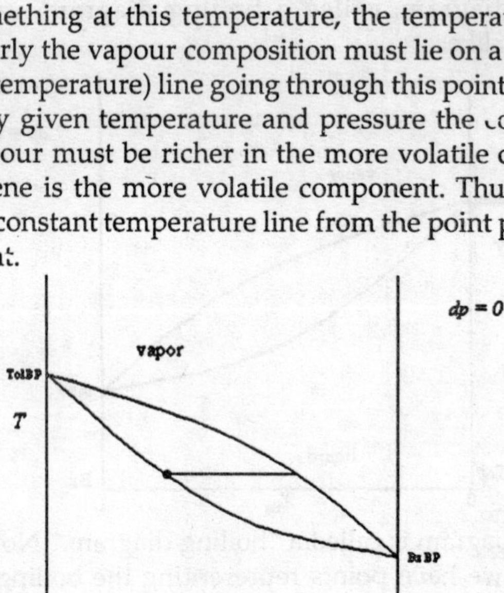

We call this line a tie line and it intersects the upper curve at the composition of the vapour in equilibrium with the liquid at the same temperature. The region between the two curves is a two-phase region. In this region two phases, liquid and gas, are in equilibrium with each other and we must be able to keep track of the compositions of both phases.

Above the upper curve the sample has fully evaporated and the system is entirely a gas. Below the lowest curve none of the sample has evaporated and the system is entirely a liquid. Between the two curved lines the system is composed of two phases in equilibrium and we must draw the tie line to keep track of the composition of both phases.

If we heat a liquid mixture at a composition of about 0.33 mole fraction benzene the liquid will have the composition shown at point "a" when the liquid first begins to boil.

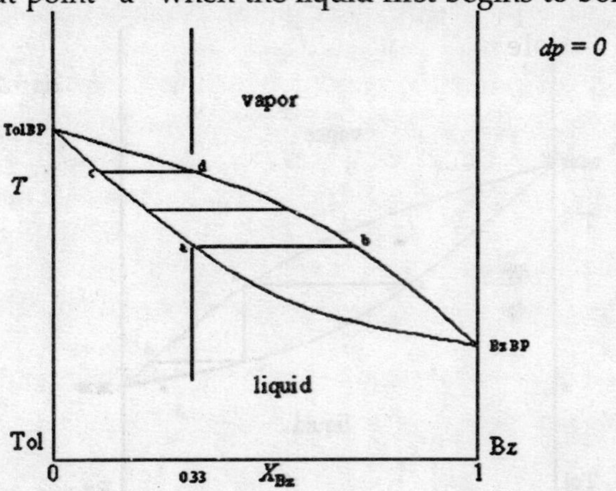

The first vapour to come off will have the composition shown at point "b." As we continue to heat the system the more volatile component will evaporate preferentially and the liquid phase will become richer in the less volatile component (toluene, here). As we continue to heat the mixture the toluene will gradually catch up to the benzene in the vapour phase. The very last bit of liquid to evaporate will have the composition shown at point "c," and the vapour will have the original composition that we started with, "d" (or "a"). Notice that as we heat the system the tie line "tracks" with the lower and upper curves to give the composition of liquid and vapour in equilibrium.

FRACTIONAL DISTILLATION

Boiling diagrams provide an explanation for fractional distillation. Suppose we start with a liquid mixture with 0.33 mole fraction benzene. As we heat the liquid it will begin to boil when the temperature reaches the temperature of point "a." The first vapour to come off has the composition shown at point "b." Capture the vapour, condense it, and heat it up. The new liquid will boil at point "c" giving a vapour with composition at point "d." Capture this vapour, cool it, and boil it at point, "e," and so on. by continuing this process the vapour can be made as pure in benzene as desired.

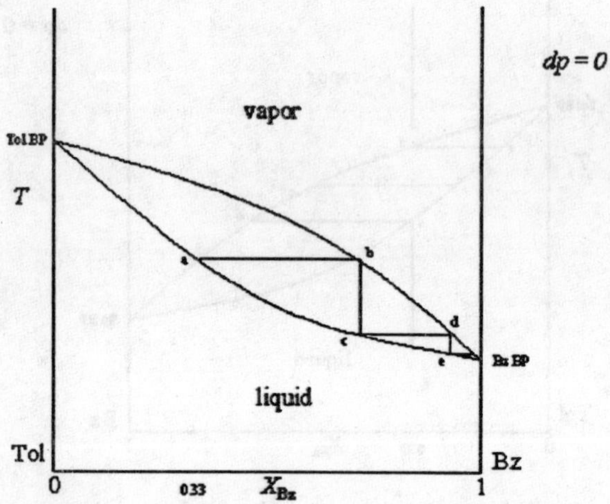

This is what fractional distillation does. There are more modern ways to separate mixtures, for example, chromatography, but fractional distillation still has industrial and some laboratory importance.

BOILING DIAGRAMS FOR NONIDEAL SOLUTIONS

We saw above that a solution with positive deviations

from Raoult's law has a higher vapour pressure than would be given by Raoult's law. That means that the solution would boil at a lower temperature than would be expected from Raoult's law. A boiling diagram for such a substance might look like the following: Point "a" denotes a constant boiling mixture called an "azeotrope." In this case it is a "low boiling azeotrope." We have drawn in some "zig-zag" lines to show why fractional distillation will not give complete separation of the mixture into its components.

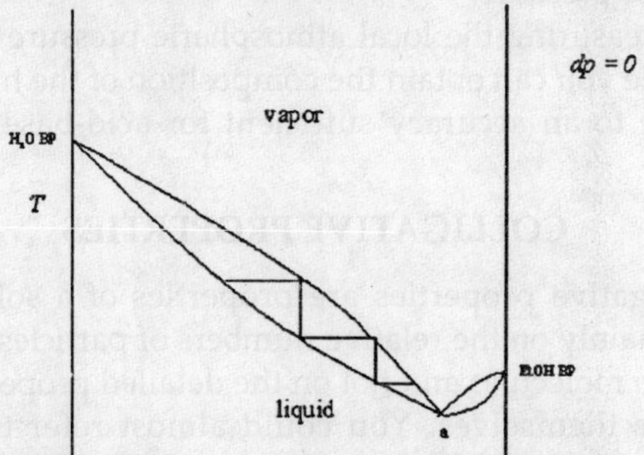

In a mixture with negative deviations from Raoult's law the vapour pressure is lower than would be expected from Raoult's law which will produce boiling points higher than expected and a high boiling azeotrope.

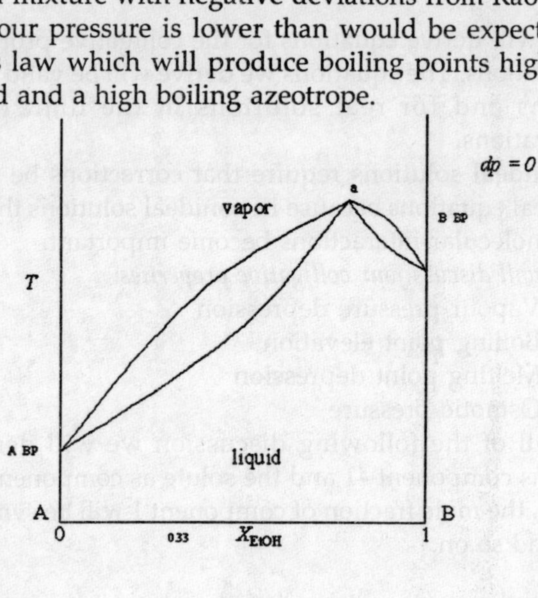

On figure the point "a" indicates the high boiling azeotrope.

It turns out that the composition of the azeotrope is a function of the pressure. One can obtain a solution with very accurately known composition by letting it boil until the high boiling azeotrope is reached.

For example, there are tables in the literature that give the composition of the high boiling HCl-H_2O azeotrope as a function of pressure.

By measuring the local atmospheric pressure and going to the table you can obtain the composition of the high boiling azeotrope to an accuracy sufficient for acid-base analytical work.

COLLIGATIVE PROPERTIES

Colligative properties are properties of a solution that depend mainly on the relative numbers of particles of solvent and solute molecules and not on the detailed properties of the molecules themselves. You could almost refer to these as statistical properties because they can be understood solely on the basis of counting the relative numbers of particles in a solution.

We will derive equations for the colligative properties of ideal solutions. The equations we derive will be valid for ideal solutions and for real solutions in the limit of small concentrations.

Nonideal solutions require that corrections be made to these ideal equations because in nonideal solutions the details of intermolecular interactions become important.

We will discuss four colligative properties:

1. Vapour pressure depression
2. Boiling point elevation
3. Melting point depression
4. Osmotic pressure

In all of the following discussion we will denote the solvent as component #1 and the solute as component #2. For example, the mole fraction of component 1 will be symbolized by X_1, and so on.

Vapour Pressure Depression

Vapour pressure depression is the simplest of the colligative properties and the easiest to understand on the basis of a physical model.

A given solvent has a vapour pressure which we usually denote by p_1^*. That means that at equilibrium the gas phase above the solvent has a solvent partial pressure of p_1^*. When you add a solute to the solvent to make a solution the partial pressure of the solvent in the gas phase decreases.

We can show that this is true using Raoult's law. In this case we have X_1 close to 1 and X_2 small.

Then Raoult's law applied to the vapour pressure (partial pressure in the gas phase) of the solvent is,

$$p_1 = X_1 p_1^* = (1 /0 X_2) p_1^*, \quad (1)$$

where we have taken advantage of the fact that in a two-component system, $X_1 + X_2 = 1$. Rearrange equation 1 to show the change in the vapour pressure of component 1,

$$p_1 - p_1^* = - X_2 \, p_1^*. \quad (2)$$

We see that the change in the vapour pressure is negative which means that the vapour pressure decreased and it decreased by an amount that is proportional to the vapour pressure of the pure liquid and proportional to the fraction of solute molecules in the solution (that is, the ratio of the number of solute molecules to the total number of molecules).

Vapour pressure depression is relatively easy to understand on the basis of a physical model. At the surface of a liquid there is a competition between the kinetic energy of the molecules (thermal energy), which is trying to push the molecules off the surface into the gas phase, and the intermolecular forces, which are trying to keep the molecules on the surface.

We will see later that the kinetic energy depends on temperature and that the average kinetic energy is proportional to the Kelvin temperature.

However, at any given temperature there is a distribution of kinetic energies with most of the molecules having a kinetic energy near the average, but with some higher and some lower.

As you increase the temperature you get more molecules with the higher kinetic energy. Some of the molecules on the surface with sufficient kinetic energy to overcome the intermolecular forces will escape into the gas phase and contribute to the pressure of solvent molecules in the gas phase.

As this pressure increases, there are more molecules in the gas phase colliding with the surface and sticking. At equilibrium the number of molecules leaving the surface just balances those returning to the surface.

The measured pressure at this point is the vapour pressure. The solute molecules decrease the vapour pressure because some of the solvent molecules on the surface have now been replaced by solute molecules.

There are fewer solvent molecules on the surface to escape and the vapour pressure goes down. You can calculate how much the pressure goes down by counting how many of the solvent molecules on the surface have been replaced by solute molecules.

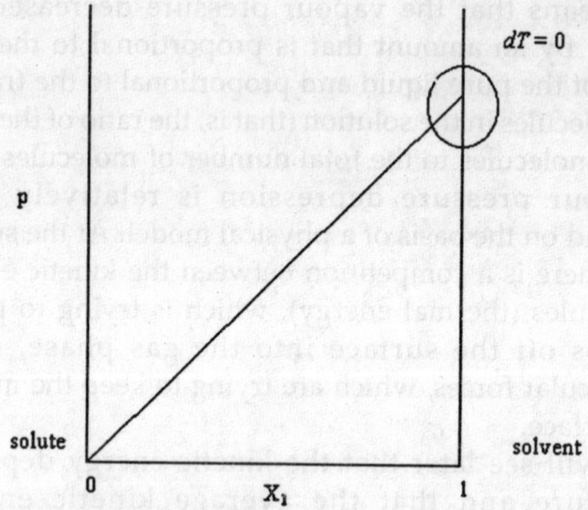

We can see graphically how the vapour pressure of the solvent decreases with the addition of solute by looking at the vapour pressure diagram below.

In this diagram the vapour pressure of the solute (component 1) is plotted against X_1 at constant temperature using Raoult's law. The circled area represents the region of small X_2 we are talking about, although for an ideal solution the straight line extends to X_1 equals zero.

The graph in the circled area would be approximately correct even for nonideal solutions because all solutions become ideal with respect to the solvent in the limit as X_1 goes to unity.

Boiling Point Elevation

It is not hard to understand qualitatively why the boiling point of a solution would be higher than the boiling point of the pure solvent.

We are assuming that the solute is a nonvolatile compound and does not make any significant contribution to the vapour pressure of the solution.

If the solute is a volatile compound with a lower boiling point than the solvent the boiling point of the solution will actually go down - by Raoult's law.

We have learned that the vapour pressure of the solvent decreases in a solution.

If our solvent was originally at the boiling point (p_1 = 1 atm) adding a nonvolatile solute will lower the vapour pressure and the mixture will no longer be boiling.

We can bring it back to boiling by increasing the temperature. Thus, the boiling point of the solution is higher that that of the pure solvent.

In the diagram below on the left side we have pure solvent at its boiling point.

We know that the chemical potential of the liquid solvent must equal the chemical potential of the solvent in the vapour phase.

That is,

$$\mu_{1l} = \mu_{1vap}. \tag{3}$$

Thus, for the vapourization process,

$$\text{liquid} \rightarrow \text{vapour}, > \tag{4}$$

we have,

$$\Delta\mu = \mu_{1vap} - \mu_{1l} = 0 \tag{5}$$

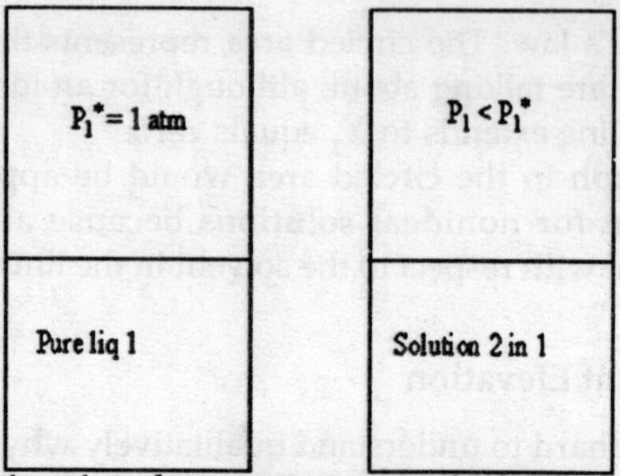

Approximating the vapour as an ideal gas and using the ideal solution equations for the solution we get.

$$\mu_{1vap} = \mu_{1vap}^{\circ} + RT\,In\,p_1 \tag{6}$$

and

$$\mu_{1l} = \mu_{1l}^{\bullet} + RT\,In X_1 = \mu_{1l}^{\bullet} + RT\,In(1 - X_2), \tag{7}$$

where, as usual, $\mu_{1l}^{\bullet}$ is the chemical potential of the pure solvent, is the standard state potential for the gas (vapour), and there is an implied 1 atm dividing p_1 inside the logarithm.

Then the difference in chemical potentials can be written,

$$\Delta\mu_1 = \mu_{1vap}^{\circ} + RT\,In\,p_1 - \mu_{1l}^{\bullet} - RT\,In(1 - X_2) \tag{8}$$

We started out, on the left of the diagram, with pure liquid solvent in equilibrium with its vapour at the normal boiling point. Let us now add a small amount of solute, component 2. Add enough to increase the mole fraction of 2 in the solution by dX_2. We insist, however, that the solvent in the solution remain in equilibrium with the solvent vapour. We now ask the question what must the temperature change, dT, be to maintain equilibrium?

$$\left(\frac{\partial T}{\partial X_2}\right)_{\Delta\mu_1} = -\frac{\left(\dfrac{\partial \Delta\mu_1}{\partial X_2}\right)_T}{\left(\dfrac{\partial \Delta\mu_1}{\partial T}\right)_{X_2}}.$$

The answer to this question is contained in the partial derivative,

$$\left(\frac{\partial T}{\partial X_2}\right)_{\Delta\mu_1} . \tag{9}$$

We can figure out how to calculate this derivative by using the cyclic relation

$$\left(\frac{\partial T}{\partial X_2}\right)_{\Delta\mu_1} = -\frac{\left(\frac{\partial \Delta\mu_1}{\partial \Delta X_2}\right)_L}{\left(\frac{\partial \Delta\mu_1}{\partial T}\right)X_2} \tag{10}$$

The upper derivative on the right is easily obtained from our expression for $\Delta\mu_1$ above and the lower derivative on the left can be obtained by recognizing that,

$$\left(\frac{\partial \mu_1}{\partial T}\right)_p = -\bar{S}_1 \tag{11}$$

or, for the vapourization process,

$$\left(\frac{\partial \Delta\mu_1}{\partial T}\right)_p = -\Delta\bar{S}_1. \tag{12}$$

Then,

$$\left(\frac{\partial T}{\partial X_2}\right)_{\Delta\mu_1} = -\frac{\left(\frac{\partial \Delta\mu_1}{\partial X_2}\right)_T}{\left(\frac{\partial \Delta\mu_1}{\partial T}\right)_{X_2}} = \frac{-RT\frac{1}{1-X_2}(-1)}{-\Delta\bar{S}_{1vap}} \tag{13}$$

We must simplify this expression, but we note first that the above expression requires no specification of the value of $\Delta\mu_1$ other than that it be held constant during the derivative process. However, we know that in order for the solvent to be in equilibrium with its vapour we must have not only $\Delta\mu_1$ = constant, but it must be held at the particular constant value of zero. That is, $\Delta\mu_1 = 0$. In this case,

$$\Delta\mu_1 = \Delta\bar{G}_1 = \Delta\bar{H}_1 - T\Delta\bar{S}_1 = 0, \tag{14}$$

from which we conclude that,

$$T\Delta\bar{S}_{1vap} = \frac{\Delta\bar{H}_{1vap}}{T}. \tag{15}$$

(We have added the subscript "vap" to remind ourselves that we are talking about the vapourization process. With these considerations Equation 13 simplifies to,

$$\left(\frac{\partial T}{\partial X_2}\right)_{\Delta\mu_1} = \frac{RT^2}{\Delta\overline{H}_{1vap}} \frac{1}{1-X_2}. \tag{16}$$

We may need this form of the equation a little later on, but for now we are going to make some approximations. We assume that X_2 is small so that 1 " $X_2 \approx 1$, then,

$$\left(\frac{\partial T}{\partial X_2}\right)_{\Delta\mu_1} = \frac{RT^2}{\Delta\overline{H}_{1vap}}. \tag{17}$$

We can set this up to integrate as follows,

$$dT = \frac{RT^2}{\Delta\overline{H}_{1vap}} dX_2. \tag{18}$$

If we are only going to integrate over a small range then our temperature is approximately constant at the boiling point, T_{BP}, and we can write,

$$\Delta T = \frac{RT_{BP}^2}{\Delta\overline{H}_{1vap}} \Delta X_2, \tag{19}$$

but $\Delta X_2 = X_2 - 0$, since we started out at zero concentration of solute, so we get,

$$\Delta T = \frac{RT_{BP}^2}{\Delta\overline{H}_{1vap}} \Delta X_2. \tag{20}$$

This equation actually solves our problem, but it is not in the form you saw in freshman chemistry. The freshman chemistry form had molality of the solute in place of the mole fraction of solute.

To get the form you expect we must relate the mole fraction to the molality,

$$X_2 = \frac{m_2}{\frac{1000}{Mw_1} + m_2}, \tag{21}$$

where m_2 is the molality of solute. That is, m_2 is the number of moles in 1000 g of solvent. (Also, Mw_1 is the formula weight of the solvent.) Since we are assming a dilute solution the number of moles of solute is much smaller than the number of moles of solvent so we can drop the m_2 in the denominator to get,

$$X_2 \approx \frac{\frac{m_2}{1000}}{Mw_1} = \frac{Mw_1 m_2}{1000}. \tag{22}$$

Plugging this expression for mole fraction into the formula we have above for the change in the boiling point we get

$$\Delta T = \frac{RT_{BP}^2}{\Delta \overline{H}_{vap}} \frac{\frac{m_2}{1000}}{Mw_1} = \left[\frac{RT_{BP}^2}{\Delta \overline{H}_{vap}} \frac{Mw_1}{1000}\right] m_2 \tag{23}$$

The expression in square brackets is the boiling point elevation constant (or, if you like big words, the ebullioscopic constant).

If we are dealing with more concentrated solutions we can go back and integrate one of the earlier equations, like

$$dT = \frac{RT^2}{\Delta \overline{H}_{vap}} \frac{1}{1-X_2} dX_2, \tag{24}$$

or

$$\frac{dT}{T_2} = \frac{R}{\Delta \overline{H}_{vap}} \frac{dX_2}{1-X_2}. \tag{25}$$

The latter can be integrated from the boiling point of the pure liquid, T_{BP} at $X_2 = 0$ to some new boiling point, T, at concentration X_2. Integration yields,

$$-\frac{1}{T} + \frac{1}{T_{BP}} = -\frac{R}{\Delta \overline{H}_{vap}} In \frac{1}{1-X_2}. \tag{26}$$

Freezing Point Depression

Freezing point depression is done the same way we did boiling point elevation only this time we must keep the solvent in equilibrium with the pure solid rather than with the vapour. That means we consider the fusion process,

$$\text{solid} \rightarrow \text{liquid} \tag{27}$$

This time the chemical potential the solid, μ_{1s}, which is the same as the chemical potential of the pure solid solvent, μ_{1s}^*, must be the same as the chemical potential of the solvent in the solution, which we have seen before,

$$\Delta\mu_{1l} = \mu_{1l}^{\circ} + RT\, InX_1 = \mu_{1l}^{\circ} + RT\, In(1-X_2), \tag{28}$$

but this time

$$\Delta\mu_1 = \mu_{1l}^* + RT\,In(1 - X_2) - \mu_{1s}^*, \tag{29}$$

and

$$\Delta\mu_1 = 0 \tag{30}$$

if the solid and solvent are in equilibrium. Under equilibrium conditions it is true, as before, that

$$\Delta\bar{S}_{fus} = \frac{\Delta\bar{H}_{fus}}{T}. \tag{31}$$

We ask the same question as in our discussion of boiling point elevation. If we add a small amount of solute, component 2, to the solvent to produce an increase in X_2 of dX_2, and we insist that the solid stay in equilibrium with the solvent, what must the temperature change be to maintain the equilibrium. This question is answered by the same type derivative we used in our discussion of boiling point elevation, namely,

$$\left(\frac{\partial T}{\partial X_2}\right)_{\Delta\mu_1} = -\frac{\left(\dfrac{\partial\Delta\mu_1}{\partial X_2}\right)_T}{\left(\dfrac{\partial\Delta\mu_1}{\partial T}\right)_{X_2}} = \frac{RT\dfrac{1}{1-X_2}(-1)}{-\Delta\bar{S}1_{fus}}. \tag{32}$$

This reduces to

$$\left(\frac{\partial T}{\partial X_2}\right)_{\Delta\mu_1} = \frac{RT}{-\Delta\bar{H}1_{fus}}\frac{1}{1-X_2}. \tag{33}$$

Notice that we have again used the fact that $\Delta\mu_1$ is not only constant, but constant at the value zero in order to be able to write

$$\Delta\bar{S}1_{fus} = \frac{\Delta\bar{H}1_{fus}}{T}. \tag{34}$$

Notice also that this time the derivative in Equation 33 is negative. (You might want to go back and see why we got a negative sign here when we had a positive sign in boiling point elevation.) Under the circumstance where the solution is very dilute, $X_2 \ll 1$, this reduces to,

$$\left(\frac{\partial T}{\partial X_2}\right)_{\Delta\mu 1} = -\frac{RT_{FP}^2}{\Delta\bar{H}1_{fus}}, \tag{35}$$

and we can integrate through a short concentration range as before to get

$$\Delta T = -\frac{RT_{FP}^2}{\Delta \overline{\overline{H}}1_{fus}} X_2. \tag{36}$$

Again, this is not in the same form you saw in freshman chemistry. To get there we have to convert from mole fraction to molality as before.

We get,

$$\Delta T = -\frac{RT_{FP}^2}{\Delta \overline{\overline{H}}1_{fus}} \frac{m_2}{\frac{1000}{Mw_1}} = -\left[\frac{RT_{FP}^2}{\Delta \overline{\overline{H}}1_{fus}} - \frac{Mw_1}{1000}\right] m_2. \tag{37}$$

The part in brackets is the freezing point depression constant or the "cryoscopic" constant. If we want a larger concentration range we have to go back and integrate,

$$\left(\frac{\partial T}{\partial X_2}\right)_{\Delta\mu} = -\frac{RT}{-\Delta \overline{\overline{H}}1_{fus}} \frac{1}{1 - X_2}, \tag{38}$$

to get,

$$-\frac{1}{T} + \frac{1}{T_{FP}} = +\frac{RT}{\Delta \overline{\overline{H}}1_{fus}} IN \frac{1 - X_2}{1}. \tag{39}$$

This expression will be of use later when we talk about solid-liquid phase diagrams.

OSMOTIC PRESSURE

The treatment of osmotic pressure is not identical to the treatments of boiling point elevation and freezing point depression we have just given. A cartoon of an osmotic pressure apparatus is shown below. A U-tube has a "semipermeable membrane inserted in the centre of the cross piece as shown. (A semipermeable membrane will allow solvent to flow through it, but not solute.) The left arm of the U-tube is filled with pure solvent and the right arm contains solution. The chemical potential of the solvent is higher in the pure solvent than in the solution so that solvent will flow through the semipermeable membrane from the pure solvent side into the solution side unless it is restrained. (Another way to look at this is that the solution "would like to become more dilute" and it will do so if it can. Given the chance, material will flow from a region of higher chemical potential to a region of lower chemical potential.) To prevent the solvent from flowing we apply an external pressure, Π, to the right side. The pressure, Π, which exactly stops the flow is called the osmotic pressure.

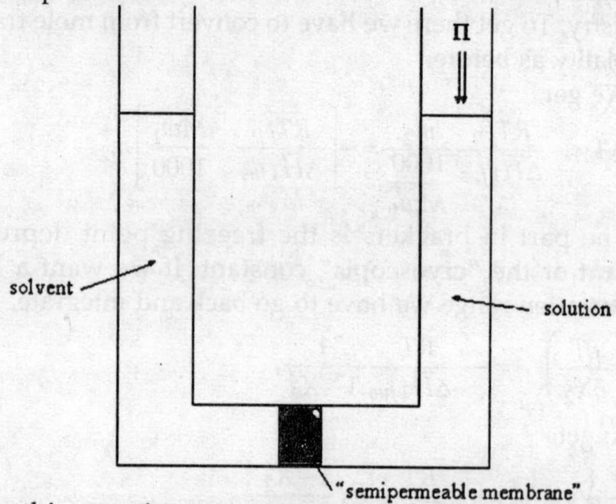

In this experiment temperature is held constant. We will bring the solvent on the solution side into equilibrium with the pure solvent by the exertion of an external pressure on the solution. As we add solute we lower the chemical potential of the solvent in the solution so we must increase the pressure on the solution to bring the chemical potential back up. That is, as we make the solution we need to increase the external pressure to keep μ_{1l} constant. Recall that

$$\mu_{1l} = \mu_{1l}^{*} + RT\, In X_1 = \mu_{1l}^{*} + RT\, In(1 - X_2). \tag{40}$$

The question we ask is: if we increase the concentration of solute by an amount dX_2, how much does the external pressure have to change to keep the solution in equilibrium with the pure solvent? The derivative we want is,

$$\left(\frac{\partial T}{\partial X_2}\right)_{\mu_1} = -\frac{\left(\dfrac{\partial \mu_1}{\partial X_2}\right)_{p,T}}{\left(\dfrac{\partial \Delta\mu_1}{\partial T}\right)_{X_2,T}} = \frac{RT\dfrac{1}{1-X_2}(-1)}{\overline{\overline{V}}_1}, \tag{41}$$

where we have used the fact that

$$\left(\frac{\partial \mu_1}{\partial p}\right)_T = \left(\frac{\partial \overline{G}_1}{\partial p}\right)_T = \overline{V}_1 \tag{42}$$

Then Equation 41 simplifies to,

$$\left(\frac{\partial p}{\partial X_2}\right)_{\mu_1} = \frac{RT}{\overline{\overline{V}}_1}\frac{1}{1-X_2} \approx \frac{RT}{\overline{\overline{V}}_1}. \tag{43}$$

Prepare to integrate this equation,

$$dp = \frac{RT}{\overline{\overline{V}}_1}dX_2, \tag{44}$$

which integrates to (calling the applied external pressure, or the osmotic pressure, Π),

$$\Pi - 0 = \frac{RT}{\overline{\overline{V}}_1}(X_2 - 0), \tag{45}$$

or, when the solution is dilute,

$$\Pi = \frac{RT}{\overline{\overline{V}}_1}X_2 = \frac{RT}{\overline{\overline{V}}_1}\frac{n_2}{n_1 + n_2} \approx \frac{RT}{n_1\overline{\overline{V}}_1}n_2. \tag{46}$$

But so,

$$\Pi = RT\frac{n_2}{V}. \tag{47}$$

This looks exactly like the ideal gas equation of state, except the pressure is the osmotic pressure, Π, and the system is a solution, not a gas.

This derivation of osmotic pressure assumed the solution is ideal. For ionic solutions, which are not ideal, you can get an approximate osmotic pressure by taking account of the dissociation of the compound into ions in solution. If the compound gives c ions per compound unit ("molecule") the osmotic pressure for dilute ionic solutions can be written,

$$\Pi \approx cRT\frac{n_2}{V}. \tag{48}$$

For more accurate work you can write an expansion similar to the virial expansion,

$$\frac{\Pi V}{n_2RT} = 1 + B\frac{n_2}{V} + C\left(\frac{n_2}{V}\right) + \ldots, \tag{49}$$

or

$$\frac{\Pi V}{cn_2 RT} = 1 + B\frac{n_2}{V} + C\left(\frac{n_2}{V}\right)^2 + \ldots, \tag{50}$$

for ionic solutions.

Osmotic pressure is enormously important for life. For example, cell wall membranes, in many cases, are semipermeable membranes which will pass molecules of water but not solu^te molecules or ions. As a consequence the concentrations of material in the fluids inside and outside the cell must be carefully balanced so that they generate the same osmotic pressure.

Otherwise, the difference in solute concentrations would generate a pressure differential across the cell wall and either collapse the cell or burst it. Fluids used in intravenous feeding, drug delivery, or just plain fluid replacement are called isotonic solutions because they are designed to have the same osmotic pressure as normal blood cells. The typical isotonic solution is 8.9 g NaCl per liter of water. If a blood cell is placed in a solution with a lower concentration than this (a hypotonic solution) fluid moves from the solution into the cell and cell bursts (called hemolysis). It is instructive to calculate the osmotic pressure of an isotonic solution (8.9g NaCl per liter of water). This is the pressure that would be generated inside a blood cell that was placed in pure water. Do you suppose that the cell wall could withstand this pressure?

GIBBS PHASE RULE

Let's try to motivate the Gibbs phase rule. Consider a typical one"component phase diagram.

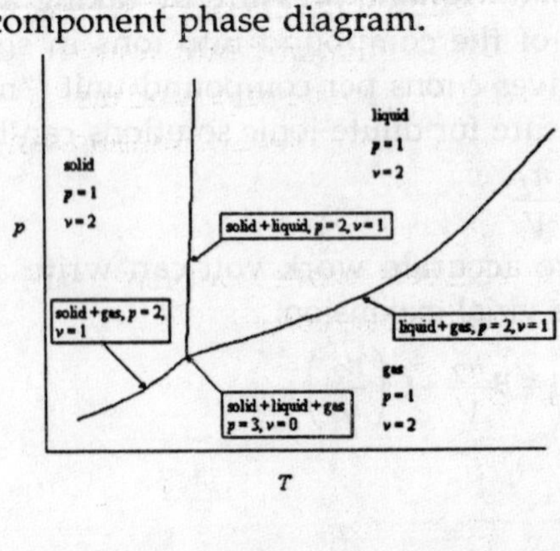

In the one"phase regions one can vary either the temperature, or the pressure, or both (within limits) without crossing a phase line. We say that in these regions there is a variance of 2. We have indicated in the solid, liquid, and gas regions that there is one phase and the variance is two. Along a phase line we have two phases in equilibrium with each other, so on a phase line the number of phases is 2. However, if we want to stay on a phase line, we can't change the temperature and pressure arbitrarily.

If we change the temperature " and keep two phases " then the pressure must change also to keep us on the phase line. Otherwise we go off the line and we no longer have two phases in equilibrium. So on a phase line the number of phases is 2, but the variance is 1. At the triple point there are three phases in equilibrium, but there is only one point on the diagram where we can have three phases in equilibrium with each other. Therefore, at the triple point the variance is zero. Notice that the variance seems to be related to the number of phases, such as,

$v = 3 - p$, or (1)

$v = 2 + 1 - p$. (2)

(You will see why we wrote it this way in a minute.)

Now let's look at another example. Consider the boiling diagram of a two"component ideal liquid"liquid solution.

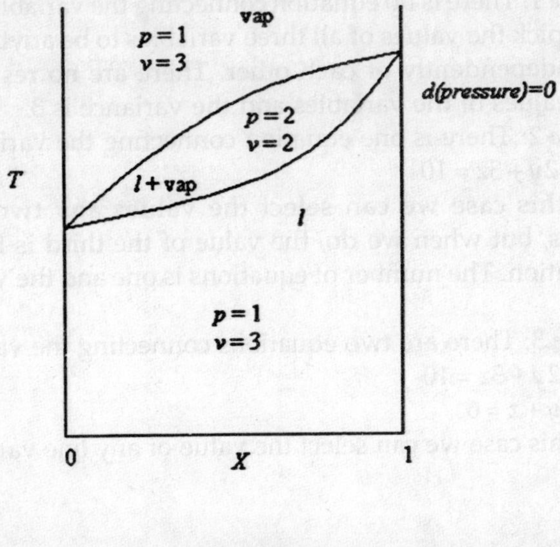

In the one"phase regions (liquid or vapour) we have a variance of three because we can change the temperature, the composition, X, and the pressure (by moving in and out of the plane of the screen). In the two phase region the variance is only two because if we change the temperature the compositions of both the liquid and vapour phases must track as indicated by the tie"lines connecting the liquid and vapour composition lines. Here it looks like the variance satisfies,

$$v = 4 - p, \text{ or} \tag{3}$$

$$v = 2 + 2 - p. \tag{4}$$

Notice that equations 2 and 4 are the same except for the second term. In Equation 2 there is one component and the second term is 1. In Equation 4 there are two components and the second term in Equation 4 is a 2. It looks like the second term may indicate the number of components in the system. We might guess that the general form for Equations 2 and 4 might be,

$$v = 2 + c - p.$$

We shall see below whether or not our guess is correct.

Before we try to derive a general form for the variance of a system in equilibrium we need to take a look as some mathematical principles.

Take a system where we have three variables, x, y, and z. Let's look at five cases:

Case 1: There is no equation connecting the variables. Then we can pick the values of all three variables to be anything we wish, independently of each other. There are no restrictions on the values of the variables and the variance is 3.

Case 2: There is one equation connecting the variables,

$$x + 2y + 3z = 10 \tag{5}$$

In this case we can select the values any two of the variables, but when we do, the value of the third is fixed by the equation. The number of equations is one and the variance is 2.

Case 3: There are two equations connecting the variables,

$$x - 2y + 3z = 10$$

$$x + y + z = 6. \tag{6a, b}$$

In this case we can select the value of any one variable at

random, but then the values of the other two are fixed by the two equations. (Two equations in two unknowns is a soluble problem.)

Case 4: There are three equations connecting the variables,

$$x - 2y + 3z = 10$$
$$x + y + z = 6$$
$$x + 3y - z = 8. \qquad (7a, b, c)$$

In this case we can not select any of the variables arbitrarily. The values of the variables are fixed because a system of three unknowns and three (linearly independent) equations has a unique solution.

$$x = 3$$
$$y = 2$$
$$z = 1 \qquad (8a, b, c)$$

The number of equations is three and the variance is zero.

Case 5: There are four linearly independent equations connecting the variables.

$$x - 2y + 3z = 10$$
$$x + y + z = 6$$
$$x + 3y - z = 8$$
$$x + 2y + 2z = 7. \qquad (9,a, b, c, d)$$

This system does not have a solution. There are no values of the three variables which will satisfy all four equations. We conclude that we cannot have a variance less than zero.

The principle we extract here is that variance = number of variables – number of equations ≥ 0. (10)

Let us now apply this principle to phase equilibrium. Consider a system which is composed of c components. That is,

c = number of components. (11)

(We define the number of components as the minimum number of independent substances (compounds and/or elements) required to form the system. If you can make one substance out of the others you can't count it! For example, a system containing CO_2, CaO, and $CaCO_3$ has two components because you can form the $CaCO_3$ by reacting the other two. There is, however, a special case. If our system contains exactly the same number of moles of CaO and CO_2 it is a one component system because it can be made starting with just pure $CaCO_3$.

That is, you can heat the $CaCO_3$ to generate the other two in equal molar amounts.)

Number the components 1, 2, 3... *c*.

Let

p = number of phases. (12)

(We will have to be careful to distinguish between the number of phases and the pressure since we are using the same symbol.) Enumerate the phases by lower case Greek letters, $\alpha, \beta, \gamma, \delta... \zeta$. Let's now count the number of variables and the number of equations. We will get the variance from

v = variables – equations. (13)

Our variables are T, p, (pressure) and the mole fractions of the components in each phase.

$$\begin{array}{llll} T, & p, & & \\ X_{1\alpha} & X_{1\beta} & X_{1\gamma} & \ldots \\ X_{2\alpha} & X_{2\beta} & X_{2\gamma} & \ldots \\ - & - & - & - \\ - & - & - & - \\ - & - & - & - \end{array} \qquad (14)$$

Notice that there are p columns and c rows in the array of mole fraction variables. That is, there are cp mole fraction variables. This gives a total of $2 + cp$ variables.

Now let's count the equations. First of all the mole fractions in each phase must sum to unity.

That is,

$$\begin{aligned} X_{1\alpha} + X_{2\alpha} + X_{2\alpha} + \ldots\ldots\ldots &= 1 \\ X_{1\beta} + X_{2\beta} + X_{2\beta} + \ldots\ldots\ldots &= 1 \end{aligned} \qquad (15)$$

Since there is one such equation for each phase this gives p equations. Next, the chemical potential of each component must be the same in every phase. (In the following we designate ζ as the last phase, that is the phases are denoted α, β, γ..., ζ.)

$$\begin{array}{llll} \mu_{1\alpha} = \mu_{1\beta} & \mu_{2\alpha} = \mu_{2\beta} & \ldots & \mu_{c\alpha} = \mu_{c\beta} \\ \mu_{1\beta} = \mu_{1\gamma} & \mu_{2\beta} = \mu_{2\gamma} & \ldots & \mu_{c\beta} = \mu_{c\gamma} \\ & - & - & - \quad - \end{array}$$

$$\bar{\mu}_{1(\zeta-1)} = \bar{\mu}_{1\zeta} \quad \bar{\mu}_{2(\zeta-1)} = \bar{\mu}_{2\zeta} \qquad \bar{c}_{(\zeta-1)} = \bar{\mu}_{c\zeta} \quad (16)$$

Note that there is no row of the form $\mu_{1\zeta} = \mu_{1\alpha'}$ etc. Equations of this form are not linearly independent because they can be deduced from the other $p - 1$ equations. So for this array of chemical potential equations there are c columns and p " 1 rows, which yields $c(p$ " 1) equations. The total number of equations is then $p + c(p$ " 1). If, as we have argued above, the variance is given by,

v = variables – equations, (13)

we get,

$v = 2 + cp - p - c(p - 1)$ (16)

$v = 2 + c - p.$ (17)

Equation 17 is called the Gibbs phase rule and it is a generally valid thermodynamic equation. That is, it contains no approximations. If you go back and check it against the (motivational) examples we started with you will see that it agrees exactly with our initial observations.

The Gibbs phase rule is an important tool in the analysis of systems with several phases and one or more components. It is used not only in chemistry, but also in geology and geophysics

TWO-COMPONENT PHASE DIAGRAMS

We have previously looked at two-component boiling diagrams. These diagrams are descriptions of the state of the system on a graph of temperature versus composition (at constant pressure). At high temperatures the system is all in the gas (or vapour) phase.

At the lower temperatures the system is in the liquid phase. In between these two situations there is a region where there are two phases (vapour and liquid) in equilibrium with each other. If we continue to cool the system we will eventually reach a temperature where one or both of the pure components will freeze. At temperatures at and below the melting points the phase diagram will look something like the following hypothetical phase diagram involving a substance "A" and a substance "B." In this diagram we are plotting temperature versus the mole fraction of substance B.

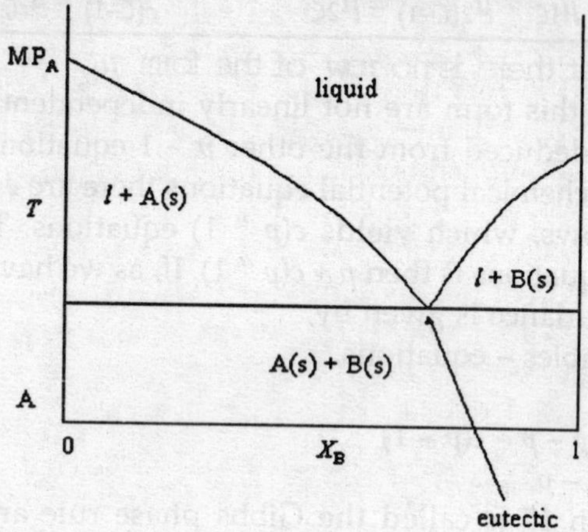

Note that the freezing point of the solution decreases as we move away from either pure A or pure B. This is a phenomenon that we have discussed previously when we talked about freezing point depression in the section on colligative properties.

Let's pick a point in the liquid region on the above diagram and cool the system at constant composition and pressure along the dotted line starting at point "a."

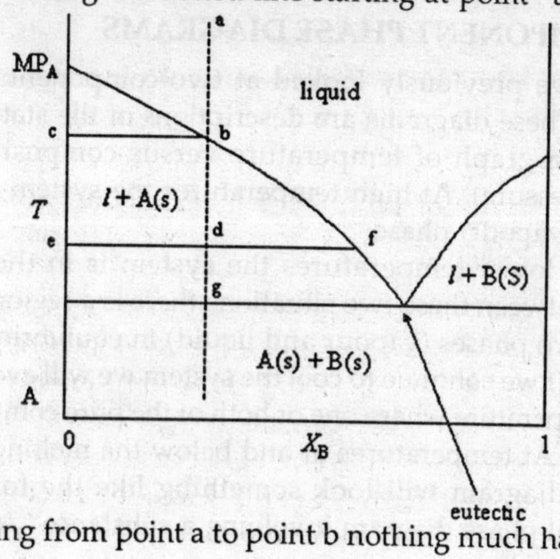

Going from point a to point b nothing much happens. We

are just cooling the liquid (or melt). At point b, however, we begin to crystallize out some pure component A. When we cross the curved line at point b we are moving into a two phase region. On this diagram the two phases are pure component A and the liquid mixture.

As we already know, in a two phase region we must track two different compositions so we draw a tie-line. In this case the tie-line runs from point b to point c.

Where the tie-line intersects the closed curve at b indicates the composition of the liquid phase. In this diagram the other end of the tie-line intersect the edge of the diagram at pure solid A (point c).

As we continue to cool the system the tie-lines track the composition of the liquid and solid phases. By the time we get to point d the left end of the tie-line (point e) still tracks the pure solid A. The right end of the tie-line tracks the liquid composition at point f. Notice that the liquid at point f is much richer in component B. This is because we have been removing component A from the solution by crystallizing it out.

When we reach point g the liquid has reached the eutectic composition. At this temperature and liquid composition both pure solid A and pure solid B will crystallize out together. In this phase diagram crystals of the two substances will be interspersed among each other. (The crystals may very well be microscopic so that the mixture looks homogeneous to the naked eye. In this case, the only way to tell that we have a mixture of two different crystals is to examine the solid under a microscope.) If we continue to remove heat from the mixture the system will remain at the eutectic temperature until all of the remaining liquid has solidified.

As the temperature decreases below point g the system crosses over into another two phase region. This time the two phases in equilibrium are pure solid A and pure solid B. Tie-lines in this region (which we have not drawn) run from the pure A side to the pure B side.

The Gibbs phase rule tells us the variance in the various regions. Since this is a two-component system the variance is given by $v = 2 + c - p = 4 - p$. One of the degrees of freedom contributing to the variance is pressure, but since pressure is constant in this diagram we have to regard pressure as being plotted perpendicular to the plane of the screen. It is usual to define the "reduced variance," v', as just the variance in the plane of the diagram. That is, the reduced variance is the phase rule variance minus the pressure variable. The reduced variance is given in a two-component phase diagram by $v' = 3$ " p. In the above diagram the reduced variance is 2 in the liquid region. (You can change temperature and composition arbitrarily over a reasonable range.) In the two-phase regions, like the A(s) + liquid, or B(s) + liquid, regions the reduced variance is 1. In these regions if you change the temperature the composition of the liquid phase must change. (Although the composition of the solid phase will not change in these cases because the solid phase is pure A or pure B.)

SOLID/SOLID SOLUBILITY

In many cases when we cool the melt to its "freezing point" it is not one of the pure components that crystallizes out, but a solid solution. A solid solution has many of the characteristics of a liquid solution except that it is solid. The primary similarity is that in a solid solution the "solute" is distributed homogeneously throughout the "solvent" as atoms or molecules (not as microcrystals). On the diagram below the region labeled α(s) is a solid solution of B in A. That is, component A would be regarded as the solvent and B as the solute. Some people would refer to this phase as "impure A." Likewise, the region labeled β(s) is a solid solution of A in B. It could also be regarded as impure B.

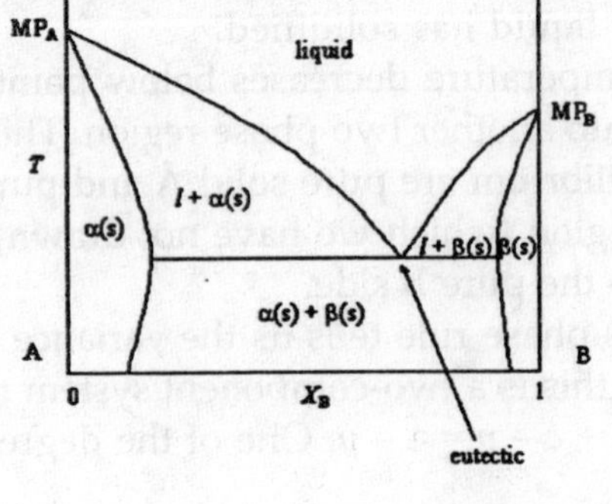

The reduced variance in the solid solution regions labeled α(s) and β(s) is 2 because these regions are one-phase regions. The reduced variance in the other regions is the same as in the phase diagram with no solid-solid solubility. On this diagram it is easier to see why the reduced variance is 1 in the region labeled α(s) + β(s). The composition of the two solid solution phases in equilibrium changes slightly as the temperature is changed. Silver and copper form solid-solid solutions similar to the above diagram where the solid-solid solubility of one component in the other is not very large.

Diffusion in solids is very slow at room temperature. Diffusion is much faster at temperatures near the melting points of the components. There is a sculpture in a Manhattan park entitled *3000* A. D. *Aluminum/Magnesium* by Terry Fugate-Wilcox. It is a spire of alternating pieces of aluminum and magnesium. Presumably solid-solid diffusion will transform the sculpture into a homogeneous solid-solid solution of aluminum and magnesium by the year 3000.

COMPOUND FORMATION

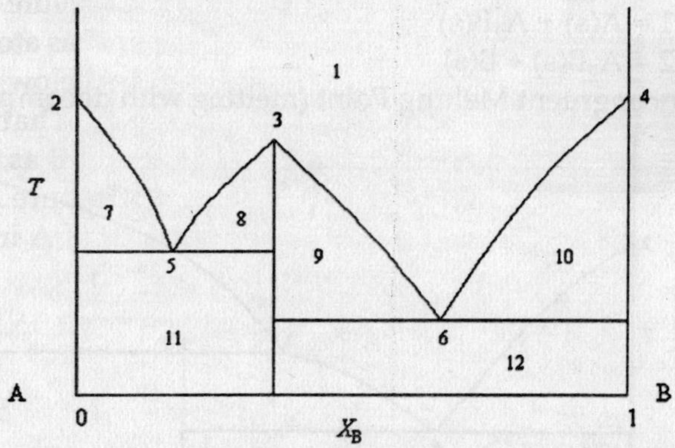

The above phase diagram shows the formation of a compound between A and B. The compound is shown by the vertical line from point 3. You can tell the formula of the compound from its composition, X_B. It looks like the compound has a mole fraction of B of 1/3. That means that in the compound,

$$X_B = \frac{n_B}{n_A + n_B} = \frac{1}{3}.$$

To find the formula of the compound we look for the smallest whole numbers for n_A and n_B which will satisfy this equation. In this case it looks like the equation is solved by setting $n_B = 1$ and $n_A = 2$.

Then,

so the formula of the compound is A_2B.

The diagram is labeled as follows:

1 = liquid
2 = MP of A
3 = MP of compound A_2B
4 = MP of B
5 = eutectic of A and A_2B
6 = eutectic of A_2B and B
7 = A(s) + liquid
8 = liquid + A_2B(s)
9 = A_2B(s) + liquid
10 = liquid + B(s)
11 = A(s) + A_2B(s)
12 = A_2B(s) + B(s)

Incongruent Melting Point (melting with decomposition).

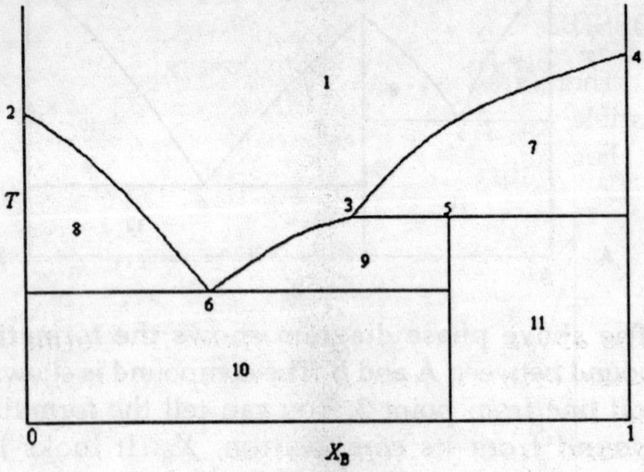

$$X_B = \frac{2}{3} = \frac{n_B}{n_A + n_B} = \frac{2}{1+2},$$

This phase diagram shows an incongruent melting point. The vertical line at point 5 represents formation of a compound.

It looks like the composition of the compound is $X_B = 2/3$, from which we conclude using the previous arguments, so that the compound is AB_2.

Point 5 is the melting point of AB_2, but notice that melting AB_2 does not give liquid of the same composition. Rather, melting of AB_2 gives liquid with the composition at point 3 and pure B(s).

So the compound, AB_2, melts and decomposes at the same time. An analysis of the points and regions is:

1 = liquid
2 = MP of A
3 = thc peritectic point (we don't get a eutectic of AB_2 and B)
4 = MP of B
5 = the incongruent melting point of AB_2
6 = eutectic of A and AB_2
7 = liquid + B(s)
8 = A(s) and liquid
9 = liquid and AB_2(s)
10 = A(s) + AB_2(s)
11 = AB_2(s) + B(s)

POSSIBILITIES

There are many possible variations and combinations possible with these themes.

For example,

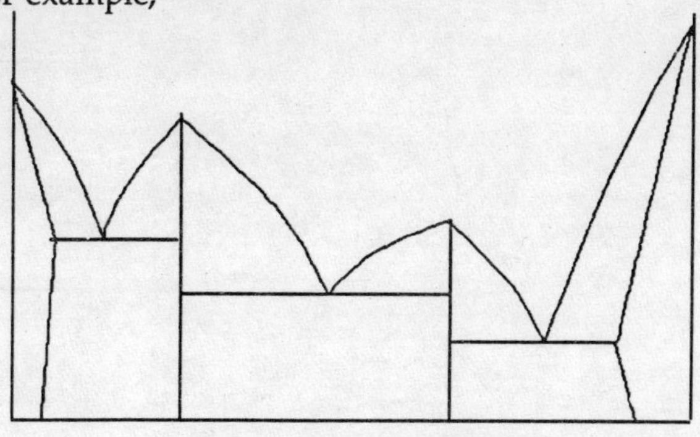

Or even more interesting,

We will leave it to the reader to understand and analyse these two hypothetical phase diagrams.

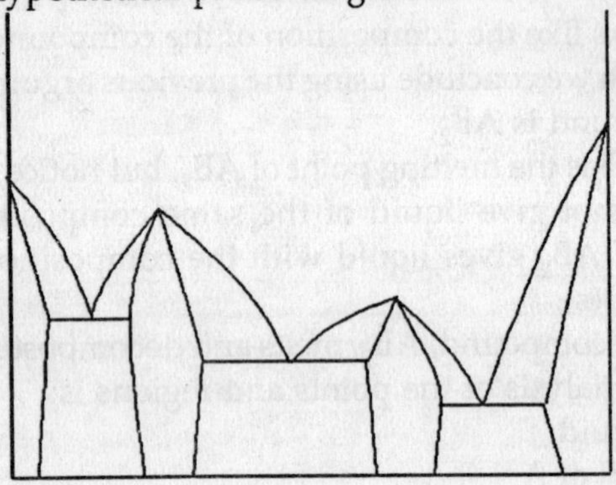

Phase diagrams, some much more complicated than the examples presented here, are very important in areas such as metallurgy, materials science, geology and geophysics, planetary science, and many more. Some of the most important ones involve more than two components.

Chapter 6

Reactions in Substances

CHEMICAL REACTIONS

A chemical reaction is material changing from a beginning mass to a resulting substance. The hallmark of a chemical reaction is that new material or materials are made, along with the disappearance of the mass that changed to make the new. This *does not mean* that new elements have been made. In order to make new elements, the nuclear contents must change. There are magnitudes of difference in the amounts of energy in ordinary chemical reactions compared to nuclear reactions, the rearrangement of the nuclei of atoms to change to new elements is enormous compared to the smaller energies of chemical changes.

The alchemists, in their efforts to change less expensive metals to gold, did not have the fundamental understanding of what they were attempting to do to appreciate the difference.

A chemical equation is a way to describe what goes on in a chemical reaction, the actual change in a material. Chemical equations are written with the symbols of materials to include elements, ionic or covalent compounds, aqueous solutions, ions, or particles. There is an arrow pointing to the right that indicates the action of the reaction.

The materials to the left of the arrow are the *reactants*, or materials that are going to react. The materials to the right of the arrow are the *products*, or materials that have been produced by the reaction. The *Law of Conservation of Mass* states that in a chemical reaction no mass is lost or gained. The Law of Conservation of Mass applies to individual types of atom. One could say that for any element, there is no loss or gain of that element in a chemical reaction. There are such things as reversible reactions, reactions in which the products reassemble to become the original products. Reversible reactions are symbolized in chemical equations by a double-headed arrow, but the standard remains to call the materials on the left the reactants and the materials on the right the products.

Chemical Reaction, process by which atoms or groups of atoms are redistributed, resulting in a change in the molecular composition of substances. An example of a chemical reaction is formation of rust (iron oxide), which is produced when oxygen in the air reacts with iron.

The products obtained from a given set of reactants, or starting materials, depend on the conditions under which a chemical reaction occurs. Careful study, however, shows that although products may vary with changing conditions, some quantities remain constant during any chemical reaction. These constant quantities, called the *conserved quantities,* include the number of each kind of atom present, the electrical charge, and the total mass.

CHEMICAL SYMBOLS

In order to discuss the nature of chemical reactions, certain basic facts about chemical symbols, nomenclature, and the writing of formulas must first be understood. All substances are made up of some combination of atoms of the chemical elements.

Rather than full names, scientists identify elements with one- or two-letter symbols. Some common elements and their symbols are carbon, C; oxygen, O; nitrogen, N; hydrogen, H; chlorine, Cl; sulfur, S; magnesium, Mg; aluminum, Al; copper, Cu; silver, Ag; gold, Au; and iron, Fe.

Most chemical symbols are derived from the letters in the name of the element, most often in English, but sometimes in German, French, Latin, or Russian. The first letter of the symbol is capitalized, and the second (if any) is lowercase. Symbols for some elements known from ancient times come from earlier, usually Latin, names: for example, Cu from *cuprum* (copper), Ag from A*rgentums* (silver), Au from *aurum* (gold), and Fe from *ferrum* (iron). The same set of symbols in referring to chemicals is used universally. The symbols are written in Roman letters regardless of language.

Symbols for the elements may be used merely as abbreviations for the name of the element, but they are used more commonly in formulas and equations to represent a fixed relative quantity of the element. Often the symbol stands for one atom of the element. Atoms, however, have fixed relative weights, called atomic weights, so the symbols often stand for one atomic weight of the element.

The atomic weights (atomic wt.) of the elements are average atomic weights of the elements as they occur in nature. Every chemical element consists of atoms the weights of which vary because of varying numbers of neutrons in their nuclei. Atoms of the same element that differ in weight are called *isotopes* of the element.

An isotope's weight may be indicated by a superscript to the left of the abbreviation that indicates the total number of nucleons (protons plus neutrons) in the nucleus. The symbols ^{235}U and ^{238}U, for example, represent two uranium isotopes of weight 235 and 238. The symbols ^{1}H, ^{2}H, and ^{3}H represent three hydrogen isotopes of weights 1, 2, and 3. If no isotopic weight is indicated, the mean (weighted average) atomic weight is indicated. All of these weights are in atomic mass units (amu). One amu is defined as u of the mass of a ^{12}C atom, the most common isotope of carbon.

An electrically neutral atom has equal numbers of protons and electrons. Electrically charged atoms and groups of atoms are called ions. When an atom is electrically charged that is, when it has lost or gained one or more electrons, and thereby become an ion that state may be indicated by a superscript to the right of the symbol, as in H^+, Mg^{++}, or Cl^-. The symbol H^+ indicates a singly positive hydrogen ion, Mg^{++} a doubly positive magnesium ion, and Cl^- a singly negative chlorine ion. The atomic number of an element is equal to the number of protons in the nucleus of an atom of the element. All isotopes of a particular element have the same number of protons in their nuclei.

The atomic number is sometimes indicated by a lower-left subscript. The symbol °U^{3+} represents a uranium ion of triply positive charge (that is, an atom that has lost 3 electrons), with 92 protons and 146 neutrons (238 nucleons – 92 protons = 146 neutrons) in its nucleus, which is surrounded by 89 electrons (92 – 3 = 89).

CHEMICAL FORMULAS

An individual atom can be represented by the symbol of the element, with the charge and mass of the atom indicated when appropriate. Most substances, however, are compound, in that they are composed of combinations of atoms. The formula for water, H_2O, indicates that two atoms of hydrogen are present for every atom of oxygen.

The formula shows that water is electrically neutral, and it also indicates (because the atomic weights are H = 1.01, O = 16.00) that 2.02 unit weights of hydrogen will combine with 16.00 unit weights of oxygen to produce 18.02 unit weights of water. Because the relative weights remain constant, the weight units can be expressed in pounds, tons, kilograms, or any other unit so long as each weight is expressed in the same unit as the other two.

Similarly, the formula for carbon dioxide is CO_2; for gasoline, C_8H_{18}; for oxygen, O_2; and for candle wax, CH_2. The subscripts in each case (with a 1 understood if no subscript is given) show the relative number of atoms of each element in the substance. CO_2 has 1 C for every 2 Os, and CH_2 has 1 C for every 2 Hs.

But why write O_2 and C_8H_{18} rather than simply O and C_4H_9, which show the same atomic and weight ratios? Experiments show that atmospheric oxygen consists not of single atoms (O) but of molecules made up of pairs of atoms (O_2); molecules of gasoline consist of carbon and hydrogen ratios of C_8 and H_{18} rather than any other combinations of carbon atoms and hydrogen atoms. The formulas of atmospheric oxygen and gasoline are examples of *molecular formulas.* Water consists of H_2O molecules, and carbon dioxide consists of CO_2 molecules.

Thus, H_2O and CO_2 are molecular formulas. Candle wax (CH_2), on the other hand, is not made up of molecules each containing 1 carbon atom and 2 hydrogen atoms. It actually consists of very long chains of carbon atoms, with most of the carbon atoms bonded to 2 hydrogen atoms in addition to being bonded to 2 neighbouring carbon atoms in the chain. Such formulas, which give the correct relative atomic composition but do not give the molecular formula, are called *empirical formulas.*

All formulas that are multiples of simpler ratios can be assumed to represent molecules: The formulas N_2, H_2, H_2O_2, and C_2H_6 represent nitrogen gas, hydrogen gas, hydrogen peroxide, and ethane.

However, formulas that show the simplest possible atomic ratios must be assumed to be empirical unless evidence exists to the contrary. The formulas NaCl and Fe_2O_3, for example, are empirical; the former represents sodium chloride (table salt) and the latter iron oxide (rust), but no single molecules of NaCl or Fe_2O_3 are present.

INORGANIC COMPOUNDS

All organic and inorganic compounds can be given systematic names based on the elementary composition and often the structure of the substance.

Binary inorganic compounds contain two different elements and are written with the more metallic (more electrically positive) element first. Such compounds are named by taking the name of the first element followed by the main part of the name of the second, more negative, element combined with the suffix *–ide:* NaCl, sodium chloride; CaS, calcium sulfide; MgO, magnesium oxide; SiN, silicon nitride.

When the atomic ratio differs from 1:1, a prefix to the name often makes this clear: CS_2 carbon disulfide; $GeCl_4$, germanium tetrachloride; SF_6, sulfur hexafluoride; NO_2, nitrogen dioxide;

N_2O_4, dinitrogen tetraoxide. Many groups of elements occur so often as ions that they are given names: nitrate, NO_3^-; sulfate, SO_4^{2-}; and phosphate, PO_4^{3-}. The suffix *–ate* usually indicates the presence of oxygen. The positive ion, NH_4^+, is called ammonium, as in NH_4Cl, ammonium chloride, or $(NH_4)_3PO_4$, ammonium phosphate.

Rules for naming more complicated compounds exist, but many compounds have been given *trivial* names—for example, $Na_2B_4O_7 \cdot 10\ H_2O$, borax—or *proprietary* names—$F(CF_2)_nF$, Teflon. These nonsystematic names may be convenient in some usages but they are often difficult to interpret. The accompanying table lists names and formulas of the most common polyatomic inorganic ions. They form compounds by combining in such a way that the net charge for the entire molecule is zero.

The sum of the charges on the positive ions equals the sum of the charges on the negative ions. When formed from water solutions, the compounds (termed *hydrates*) often contain water molecules, as does borax, the systematic name of which is disodium tetraborate decahydrate a good example of the advantages and disadvantages of trivial names.

Name	Formula
Acetate	$CH_3CO_2^-$
Ammonium	NH_4^+
Carbonate	CO_3^{2-}
Chromate	CrO_4^{2-}
Dichromate	$Cr_2O_7^{2-}$
Perchlorate	ClO_4^-
Chlorate	ClO_3^-
Chlorite	ClO2–
Hypochlorite	ClO–
Cyanide	CN–
Hydroxide	OH–
Permanganate	MnO_4^-
Manganate	MnO_4^{2-}
Nitrate	NO_3^-

Nitrite	NO2–
Peroxide	O_2^{2-}
Phosphate	PO43–
Silicate	SiO_4^{4-}
Sulfate	SO_4^{2-}
Sulfite	SO_3^{2-}

In the table, the suffix *–ite* indicates fewer oxygen atoms than in the corresponding –ate ion, with the prefix *hypo-* used with the suffix *–ite* indicating still fewer. The prefix *per-* indicates more oxygen, or less negative charge, than the corresponding –ate ion.

CHEMICAL EQUATIONS

Chemical symbols and formulas are used to describe chemical reactions; they denote substances having one set of formulas changing into substances having another set of formulas.

Consider the chemical reaction in which methane, or natural gas (formula CH_4), burns in oxygen (O_2), to form carbon dioxide (CO_2), and water (H_2O). If we assume that only these four substances are involved, the formulas (used mainly as abbreviations for names) would be stated:

$$CH_4 + 2O_2 \text{ give } CO_2 + H_2O$$

Because atoms are conserved in chemical reactions, however, the same numbers of atoms must appear on both sides of the equation. Therefore, the reaction might be expressed as,

$$1CH_4 + 2O_2 \text{ gives } 1\ CO_2 + 2H_2O$$

Chemists substitute an arrow for "gives" and delete all the "1's" to get the balanced chemical equation:

$$CH_4 + 2O_2 \rightarrow CO_2 + 2H_2O$$

Electrical charges and numbers of each kind of atom are conserved. Balanced chemical equations are balanced not only with respect to charge and numbers of each kind of atom but also with respect to weight, or, more correctly, to mass. The periodic table lists these atomic weights:

$$C = 12.01,\ H = 1.01,\ O = 16.00.$$

So we can identify each atomic symbol with an appropriate mass:

$$CH_4 \quad + \quad 2O_2$$

$$\underbrace{12.01 + (4 \times 1.01)}_{16.05} \quad + \quad \underbrace{2(16.00 + 16.00)}_{64.00}$$

$$\underbrace{16.05 \quad + \quad 64.00}_{80.05}$$

$$+$$

$$Co_2 \quad + \quad 2H_2O$$

$$\underbrace{12.01(2 \times 16.00)}_{44.01} \quad + \quad \underbrace{2[(2 \times 1.01) + 16.00]}_{36.04}$$

$$\underbrace{44.01 \quad + \quad 36.04}_{80.05}$$

Thus, 16.05 atomic mass units (amu) of CH_4 react with 64.00 amu of O_2 to produce 44.01 amu of CO_2 plus 36.04 amu of H_2O. Or 1 mole of methane reacts with 2 moles of oxygen to produce 1 mole of carbon dioxide plus 2 moles of water. The total mass on each side of the equation is conserved:

$$16.05 + 64.00 = 80.05 = 44.01 + 36.04$$

Thus charge, atoms, and mass are all conserved.

CHEMICAL BONDING

When two or more atoms are brought close enough, an attractive force between the electrons of individual atoms and the nuclei of one or more of the other atoms can result. If this force is large enough to keep the atoms together, a *chemical bond* is said to be formed. All chemical bonds result from the simultaneous attraction of one or more electrons by more than one nucleus.

TYPES OF BONDS

If the bonded atoms are of metallic elements, the bond is said to be *metallic*. The electrons are shared between the atoms but are able to move through the solid to give electrical and thermal conductivity, luster, malleability, and ductility.

If the bonded atoms are nonmetals and identical (as in N_2 or O_2), the electrons are shared equally between the two atoms, and the bond is called *nonpolar covalent*. If the atoms are nonmetals but differ (as in nitric oxide, NO), the electrons are shared unequally and the bond is called *polar covalent*—polar because the molecule has a positive and a negative electric pole much like the north and south poles of a magnet, and covalent because the atoms share electrons between them, even though unequally.

These substances are not electrical conductors, nor do they have luster, ductility, or malleability.

When a molecule of a substance contains atoms of both metals and nonmetals, the electrons are more strongly attracted to the nonmetals, which become negatively charged ions; the metals become positively charged ions. The ions then attract their opposites in charge, forming *ionic bonds*.

Ionic substances conduct electricity when they are in the liquid state or in water solutions, but not in the crystalline state, because individual ions are too large to move freely through the crystal.

Symmetrical sharing of electrons gives either metallic or nonpolar covalent bonds; unsymmetrical sharing gives polar covalent bonds; electron transfer gives ionic bonds. The tendency for unequal distribution of electrons between pairs of atoms generally increases as they are farther apart in the periodic table.

For the formation of stable ions and of covalent bonds, the most common pattern is for each atom to achieve the same total number of electrons as the noble gas Group 18 (or VIIIa) element closest to it in the periodic table. The metals in Groups 1 (or Ia) and 11 (or Ib) of the periodic table tend to lose one electron to form singly positive ions; those in Groups 2 (or IIa) and 12 (or IIb) tend to lose two electrons to form doubly positive ions; and similarly for Groups 3 (or IIIb) and 13 (or IIIa).

Likewise, the halogens, Group 17 (or VIIa), tend to gain one electron to form singly negative ions, and elements of Group 16 (or VIa) to form doubly negative ions. As the net charge on an ion increases, however, the ion becomes less stable with respect to sharing electrons with other atoms, so most large apparent charges (as in MnO_2, +4 and -2, respectively) would be minimized by covalent sharing of electrons.

Covalent bonds form when both atoms lack the number of electrons in the nearest noble gas atom. Neutral chlorine atoms, for example, have one less electron per atom than do krypton atoms (35 versus 36).

When two chlorine atoms form a covalent bond sharing two electrons (one from each atom), both achieve the krypton number of 36, Cl:Cl. It is common to represent a shared pair of electrons by a straight line between the atom symbols: Cl:Cl is written Cl8Cl.

Similarly, atomic nitrogen is three electrons short of the neon number (ten), but each nitrogen can get the neon number if six electrons are shared between them: N:::N or N ≡ N. This is called a triple bond.

Sulfur, in the same way, can achieve the krypton number by sharing four electrons in a double bond, S::S or S = S. In carbon dioxide, both the carbon (with six of its own electrons) and oxygen (with eight) achieve the neon number (ten) by sharing with double bonds: O = C = O. In all these bonding formulas, only the shared electrons are shown.

VALENCE

In most atoms, many of the electrons are so firmly attracted to their own nucleus that they can have no appreciable interaction with other nuclei. Only those electrons on the "outside" of an atom can interact with two or more nuclei. These are called *valence electrons*.

The number of valence electrons in an atom is indicated by the atom's periodic table family (or group) number, using only the older Roman numeral designation. Thus we have one valence electron for elements in Groups 1 (or Ia) and 11 (or Ib).

There are two valence electrons for elements in Groups 2 (or IIa) and 12 (or IIb), and four for elements in Groups 4 (or IVb) and 14 (or IVa). Each of the noble gas atoms elements except helium (that is, neon, argon, krypton, xenon, and radon) has eight valence electrons. Elements in families (groups) near the noble gases tend to react to form noble gas sets of eight valence electrons. This is known as the Lewis Rule of Eight, which was enunciated by the American chemist Gilbert N. Lewis.

The exception, helium (He), has a set of two valence electrons. Elements near helium tend to acquire a valence set of two: hydrogen by gaining one electron, lithium by losing one, and beryllium by losing two electrons. Hydrogen typically shares its single electron with one electron from another atom to form a single bond; such as in hydrogen chloride, H – Cl.

The chlorine, originally with seven valence electrons, now has eight. These valence electrons can be shown as H:C̤̈l: or H—|Cl|.

The structures of N_2 and CO_2 may now be expressed as,

:N: : N: or |N≡N|

and

:Ö::C::Ö: or ⟨O=C=O⟩.

These so-called Lewis structures show noble gas valence electron sets of eight for each atom. Probably 80 per cent of all covalent compounds can be reasonably represented by Lewis electron structures. The remainder, especially those containing elements in the central region of the periodic table, often cannot be described in terms of noble gas structures.

RESONANCE

An interesting extension of Lewis structures, called *resonance,* is found, for example, in nitrate ions, NO_3^-. Each N originally has five valence electrons, each O has six, plus one for the negative charge, or a total of 24 (5 + [3 × 6] + 1 = 24) electrons for four atoms. This is only an average of six electrons per atom, so covalent sharing must occur if the Lewis Rule of Eight is to apply.

It is known that the nitrogen atom takes a central position surrounded by the three oxygen atoms, which can give an acceptable Lewis structure, except that there are three possible structures.

Actually only one structure is observed. Each Lewis

resonance structure suggests that two bonds should be single and one double. Experiments have shown, however, that all the bonds are actually identical in every respect, with properties intermediate between those observed for single and double bonds in other compounds.

Modern theory suggests that a structure of localized, Lewis-type, shared electron bonds gives the general shape and symmetry of the molecule plus a set of delocalized electrons (shown by dotted lines) that are shared over the whole molecule.

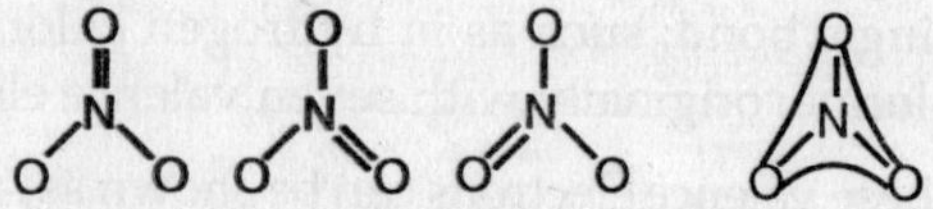

Resonance structure Observed

TYPES OF CHEMICAL REACTIONS

An understanding of reaction mechanisms can be gained from a study of ionic and covalent bonding. One kind of reaction, ion matching, is easy to understand as due to the pairing (or dissociation) of ions to form (or dissociate) neutral ionic substances, as in $Ag^+ + Cl^- \rightleftharpoons AgCl$, or $3\,Ca^{2+} + 2\,PO_4^{3+} \rightleftharpoons Ca_3(PO_4)_2$, where the double arrow (instead of an equal sign) emphasizes the two possible directions of reaction. Covalent single bond changes in which both electrons come from (or go to) one reactant are called *acid-base reactions,* as in

$$H^+ + |\overline{\underline{Cl}}^- \rightleftharpoons H - \overline{\underline{Cl}}|$$

or

$$|NH_3 + Bf_3 \rightleftharpoons H_3N - Bf_3$$

A pair of electrons from the base enter an empty electron orbital of the acid to form the covalent bond. Covalent single bond changes in which one bonding electron comes from (or goes to) each reactant are called *free radical reactions,* as in,

$$H\cdot + \cdot H \rightleftharpoons H{-}H.$$

Sometimes reactants gain and lose electrons, as in oxidation-reduction, or *redox,* reactions: $2\,Fe^{2+} + Br_2 \rightleftharpoons 2Fe^{3+} + 2Br^-$. Thus, in an oxidation-reduction reaction, one reactant is *oxidized* (loses one or more electrons) and the other reactant is *reduced* (gains one or more electrons).

Common examples of redox reactions involving oxygen are the rusting of metals such as iron (in which case the metals are oxidized by atmospheric oxygen), combustion, and the metabolic reactions associated with respiration. An example of a redox reaction that does not involve atmospheric oxygen is the reaction that produces electricity in the lead storage battery:

$$Pb + PbO_2 + 4H^+ + 2SO_4^{2-} = 2PbSO_4 + 2H_2O.$$

The joining of two groups is also called *addition;* their separation is called *decomposition.* Multiple addition involving many identical molecules is called *polymerization.*

CHEMICAL ENERGETICS

Energy is conserved in chemical reactions. If stronger bonds form in the products than are broken in the reactants, heat is released to the surroundings, and the reaction is termed *exothermic.* If stronger bonds break than are formed, heat must be absorbed from the surroundings, and the reaction is *endothermic.* Because strong bonds are more apt to form than weak bonds, spontaneous exothermic reactions are common for example, the combustion of carbon-containing fuels with air to give CO_2 and H_2O, both of which possess strong bonds. Spontaneous endothermic reactions, however, are also well known; the dissolving of salt in water is one example.

Endothermic reactions are always associated with the spreading, or the dissociation, of molecules. This can be measured as an increase in the entropy of the system. The net effect of the tendency for strong bonds to form and the tendency of molecules and ions to spread out, or dissociate, can be measured as the change in free energy of the system. All spontaneous changes at constant pressure and temperature involve an increase in free energy, with a large increase in bond strength, or a large increase in spreading out, or both.

CHEMICAL RATES AND MECHANISMS

Some reactions, such as explosions, occur rapidly. Other

reactions, such as rusting, take place slowly. Chemical kinetics, the study of reaction rates, shows that three conditions must be met at the molecular level if a reaction is to occur: The molecules must *collide;* they must be positioned so that the reacting groups are together in a *transition state* between reactants and products; and the collision must have enough *energy* to form the transition state and convert it into products.

Fast reactions occur when these three criteria are easy to meet. If even one is difficult, however, the reaction is typically slow, even though the change in free energy permits a spontaneous reaction.

Rates of reaction increase in the presence of *catalysts,* substances that provide a new, faster reaction mechanism but are themselves regenerated so that they can continue the process. Mixtures of hydrogen and oxygen gases at room temperature do not explode. But the introduction of powdered platinum leads to an explosion as the platinum surface becomes covered with adsorbed oxygen.

The platinum atoms stretch the bonds of the O_2 molecules, weakening them and lowering the activation energy. The oxygen atoms then react rapidly with hydrogen molecules, colliding with them, forming water, and regenerating the catalyst. The steps by which a reaction occurs are called the *reaction mechanism.*

Rates of reaction can be changed not only by catalysts but also by changes in temperature and by changes in concentrations. Raising the temperature increases the rate by increasing the kinetic energy of the molecules of the reactants, thereby increasing the likelihood of transition states being achieved. Increasing the concentration can increase the reaction rate by increasing the rate of molecular collisions.

CHEMICAL EQUILIBRIUM

As a reaction proceeds, the concentration of the reactants usually decreases as they are used up.

The rate of reaction will, therefore, decrease as well. Simultaneously, the concentrations of the products increase, so it becomes more likely that they will collide with one another to reform the initial reactants. Eventually, the decreasing rate of the forward reaction becomes equal to the increasing rate of the reverse reaction, and net change ceases. At this point the system is said to be at chemical equilibrium. Forward and reverse reactions occur at equal rates.

Changes in systems at chemical equilibrium are described by Le Châtelier's principle, named after the French scientist Henri Louis Le Châtelier: Any attempt to change a system at equilibrium causes it to react so as to minimize the change.

Raising the temperature causes endothermic reactions to occur; lowering the temperature leads to exothermic reactions. Raising the pressure favours reactions that lower the volume, and vice versa. Increasing any concentration favours reactions using up the added material; decreasing any concentration favours reactions forming that material.

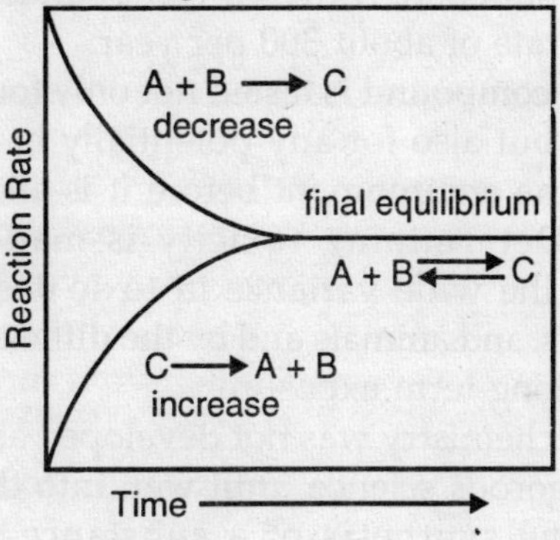

CHEMICAL SYNTHESIS

The principal goals of synthetic chemistry are to create new chemical substances and to develop better, less-expensive methods for the synthesis of known substances. Sometimes simply purifying naturally occurring substances is sufficient either to obtain an important chemical or to increase use of that chemical as a starting material for other syntheses.

For instance, the pharmaceutical industry often depends, for the source of starting materials in the synthesis of important medicines, upon the complicated organic chemicals found in crude oil. More commonly, especially for rare or expensive naturally occurring substances, it is necessary to synthesize the substance from less-expensive or more-available raw materials.

One task of synthetic chemistry, then, is to produce additional amounts of substances already found in nature. Examples are the recovery of copper metal from its ores and the syntheses of certain naturally occurring medicines (such as aspirin) and vitamins (such as ascorbic acid—vitamin C). A seccnd task is to synthesize materials not found in nature, such as steel, plastics, ceramics (space shuttle tiles, for example) and adhesives.

Some 11 million chemical compounds are now cataloged with the Chemical Abstracts Service in Columbus, Ohio; about 2000 new ones are synthesized every day. Some 6000 are in commercial production, with new compounds coming into the market at the rate of about 300 per year.

Each new compound is tested not only for its benefits and intended use, but also for any potentially harmful effects on humans and the environment before it is allowed to go into the market. Determining toxicity is made difficult and expensive by the wide variance in toxic dose levels among humans, plants, and animals and by the difficulty of measuring the effects of long-term exposure.

Synthetic chemistry was not developed as a sophisticated and highly rigorous science until well into the 20th century. Until then, the synthesis of a substance was often first accomplished by accident, and the uses of these new materials were limited. The sketchy theoretical ideas prior to the turn of the century also limited chemists' ability to develop systematic approaches to synthesis.

In contrast, it is now possible to design new chemical substances to fill specific needs, to synthesize in the laboratory almost any substance found in nature, to invent and prepare new compounds, and even to predict, based on sophisticated computer modeling, either the properties of a "target" molecule or its long-term effects in medicine or in the environment. Much of the recent progress in synthesis rests on the ability of scientists to determine the detailed structure of a range of substances and to understand the correlations between a molecule's structure and its properties, or *structure-activity relationships.* In fact, the likely structures and properties of a series of target molecules can now be modeled ahead of their synthesis, giving scientists a better understanding of the types of substances most needed for a given purpose.

Modern penicillin drugs are synthetic modifications of the substance first observed in nature by the British bacteriologist Alexander Fleming. More than 1000 human diseases have been identified as stemming from molecular deficiencies, and many can be treated by remedying that deficiency using synthetic pharmaceuticals.

Much of the search for new fuels and for methods of using solar energy is based on the study of the molecular properties of synthetic materials. One of the most recent accomplishments of this type is the fabrication of superconductors based on the structure of complicated inorganic ceramic materials, such as $YBa_2Cu_3O_7$ and other structurally similar materials.

It is now possible to synthesize hormones, enzymes, and genetic material identical to that found in living systems, thereby increasing the possibility of treating the root causes of human illness by genetic engineering. This has been made easier in recent years by computer-assisted design of syntheses and by the powerful modeling capabilities of modern computers.

One of the most successful recent developments in synthetic biochemistry has been the routine use of simple living systems, such as yeasts, bacteria, and molds, to produce important substances. The biochemical synthesis of biological materials is now possible. *Escherichia coli* bacteria, for example, are used to produce human insulin. Yeasts are also used to produce alcohol, and molds are used to produce penicillin.

REACTION TYPES

Synthesis (Composition) means that two pieces join together to produce one, more complex compound (product). Written using generic symbols, it is usually shown as: $A + B \rightarrow AB$. These pieces A and B can be elements or simpler compounds. The complex product compound has more atoms than the reactant molecules.

These are some examples (skeleton equations given - equations are not balanced):

$Mg(s) + O_2(g) \rightarrow MgO(s)$

$H_2(g) + O_2(g) \rightarrow H_2O(g)$

$K(s) + Cl_2(g) \rightarrow KCl(s)$

$Fe(s) + O_2(g) \rightarrow Fe_2O_3(s)$

$CaO(s) + CO_2(g) \rightarrow CaCO_3(s)$

$Na_2O(s) + CO_2(g) \rightarrow Na_2CO_3(s)$

$KCl(s) + O_2(g) \rightarrow KClO_3(s)$

There is only one substance on the right-hand (product) side. This is not always the case in a synthesis reaction. Sometimes there will be two products.

Here's an example:

$CO_2(g) + H_2O(l) \rightarrow C_6H_{12}O_6(aq) + O_2(g)$ You might recognize this as the photosynthesis equation.

Decomposition one compound splits apart into two (or more pieces). These pieces can be elements or simpler compounds. Written using generic symbols, it is usually shown as:

$$AB \rightarrow A + B.$$

These pieces A and B can be elements or simpler compounds.

These are some examples (skeleton equations given - equations are not balanced):

$HgO(s) \rightarrow Hg(l) + O_2(g)$

$H_2O(l) \rightarrow H_2(g) + O_2(g)$

$MgCl_2(s) \rightarrow Mg(s) + Cl_2(g)$

$FeS(s) \rightarrow Fe(s) + S(s)$

Decomposition can also split one compound into two simpler compounds (or compound and an element) as in these examples (skeleton equations given - equations are not balanced)::

$CaCO_3(s) \rightarrow CaO(s) + CO_2(g)$

$Na_2CO_3(s) \rightarrow Na_2O(s) + CO_2(g)$

$KClO_3(s) \rightarrow KCl(s) + O_2(g)$

$Ba(ClO_3)_2(s) \rightarrow BaCl_2(s) + O_2(g)$

- Binary compounds (like the first example set above) will break down into their elements.
- Carbonates (like the first two in the second example set above) break down to the metal oxide and carbon dioxide.
- Chlorates (like $KClO_3$ and $Ba(ClO_3)_2$ in the example) will break down to the binary chloride salt and oxygen.

Single Replacement (Displacement) During single replacement, one element replaces another element in a compound. There are two different possibilities: One cation replaces another.

Written using generic symbols, it is:

$$AX + B \rightarrow BX + A$$

Element B has replaced A (in the compound AX) to form a new compound BX and the free element A. Remember that A and B are both cations (postively-charged ions) in this example. These are some examples (skeleton equations given - equations are not balanced):

$Cu(s) + AgNO_3(aq) \rightarrow Ag(s) + Cu(NO_3)_2(aq)$

$Fe(s) + Cu(NO_3)_2(aq) \rightarrow Fe(NO_3)_2(aq) + Cu(s)$

$Ca(s) + H_2O(l) \rightarrow Ca(OH)_2(aq) + H_2(g)$

$Zn(s) + HCl(aq) \rightarrow ZnCl_2(aq) + H_2(g)$

Notice how, when hydrogen gets displaced, it is written as a diatomic. One anion replaces another. Written using generic symbols, it is:

$$X + AY \rightarrow AX + Y$$

Element X has replaced Y (in the compound AY) to form a new compound AX and the free element Y. Remember that X and Y are both anions (negatively-charged ions) in this example. Here are two examples (skeleton equations given - equations are not balanced)::

$Cl_2\ (g) + NaBr\ (aq) \rightarrow NaCl(aq) + Br_2\ (l)$

$Br_2\ (aq) + KI\ (aq) \rightarrow KBr\ (aq) + I_2\ (aq)$

In single replacement, one reactant is always an element. It does not matter if the element is written first or second on the reactant side. The other reactant will be a compound.

Double Replacement (Displacement)

During double replacement, the cations and anions of two different compounds switch places.

Written using generic symbols, it is:

$AX + BY \rightarrow AY + BX$

A and B are the cations (postively-charged ions) in this example, with X and Y being the anions (negatively-charged ions).

These are some examples:

$KOH(aq) + H_2SO_4(aq) \rightarrow K_2SO_4(aq) + H_2O(l)$

$FeS(s) + HCl(aq) \rightarrow FeCl_2(aq) + H_2S(g)$

$NaCl(s) + H_2SO_4(aq) \rightarrow Na_2SO_4(aq) + HCl(g)$

$AgNO_3(aq) + NaCl(aq) \rightarrow AgCl(s) + NaNO_3(aq)$

There are some special cases of double replacement reactions.

An example is

$CaCO_3(s) + HCl(aq) \rightarrow CaCl_2(aq) + H_2CO_3(aq)$ but $H_2CO_3(aq)$ is not stable and will immediately decompose as shown:

$$H_2CO_3(aq) \rightarrow H_2O(l) + CO_2(g).$$

The reaction is double replacement followed by decomposition, the reaction is best represented by:

$$CaCO_3(s) + HCl(aq) \rightarrow CaCl_2(aq) + H_2O(l) + CO_2(g)$$

More examples:

$CaCO_3(s) + HNO_3(aq) \rightarrow Ca(NO_3)_2(aq) + CO_2(g) + H_2O(l)$

$K_2SO_3(s) + HNO_3 \rightarrow KNO_3(aq) + SO_2(g) + H_2O(l)$

$NH_4Cl(aq) + NaOH(aq) \rightarrow NaCl(aq) + NH_3(g) + H_2O(l)$

Notice how, one of the two product compounds decomposes. Whenever H_2CO_3, H_2SO_3, or NH_4OH is a product formula, the correct technique is to write the products as done in the examples.

Combustion written as a general equation complete combustion is usually shown as:

$$C_xH_y + O_2 \rightarrow CO_2 + H_2O$$

These are some examples (skeleton equations given - equations are not balanced):

CH_4 (g) + O_2 (g) → CO_2 (g) + H_2O (g)
C_2H_6 (g) + O_2 (g) → CO_2 (g) + H_2O (g)
$C_6H_{12}O_6$ (g)+ O_2 (g)→ CO_2 (g)+ H_2O (g)
C_2H_5OH (l) + O_2 (g) → CO_2 (g)+ H2O (g)

Notice that some compounds contain carbon, hydrogen AND oxygen. However, the products are all the same, in every reaction. Isn't that great? We could vary it a bit by adding nitrogen (burns to form NO2) to the compound formula or sulfur (burns to form SO_2).

Like this:

$C_{21}H_{24}N_2O_4$(s) + O_2 (g) → CO_2 (g) + H_2O (g) + NO_2 (g)
C_2H_5SH(g) + O_2 (g)→ CO_2 (g) + H_2O (g) + SO_2 (g)

Combination and Decomposition Reactions

Combination reactions follow the simple formula: A + B = C, where A and B are elements or compounds that combine to form the new compound C. For example, aluminum (Al) combines rapidly with molecular oxygen (O_2) to form aluminum oxide (Al_2O_3): $4Al + 3O_2 \rightarrow 2Al_2O_3$ The numbers before each formula refer to the relative amounts of each element or compound in the reaction. Aluminum oxide forms a protective layer over aluminum products such as soda cans and foil, preventing further oxidation of the metal. In contrast, iron is poorly protected by its oxide layer and subsequently rusts with exposure to air. Another combination reaction occurs when sulfur (S) reacts with molecular oxygen to form gaseous sulfur dioxide (SO_2): $S_{(s)} + O_{2(g)} \rightarrow SO_{2(g)}$. The letter subscripts in the equation refer to the physical states of the reactants and product—*s* stands for solid, *g* stands for gas, and an *l* would stand for liquid.

A decomposition reaction is the reverse of a combination reaction. For example, when solid potassium chlorate ($KClO_3$) is heated, it decomposes into molecular oxygen and potassium chloride (KCl):

$$2KClO_{3(s)} \rightarrow 3O_{2(g)} + 2KCl_{(s)}.$$

The oxygen product of this decomposition reaction is useful for applications such as medical emergencies. Another decomposition reaction is the rapid decay when exposed to light of silver iodide (AgI) into silver (Ag) and iodine (I): 2AgI $\rightarrow$ 2Ag + I_2.

Oxidation-Reduction Reactions

Oxidation-reduction, or redox, reactions combine chemicals wanting to gain electrons (or be reduced) with chemicals that are willing to give up electrons (or be oxidized). For example, sodium (Na), with its single loosely held valence electron, gives up its outer electron (or is oxidized) by sulfur (S) to form sodium sulfide (Na_2S): 2Na + S $\rightarrow$ Na_2S. In this redox reaction, two sodium atoms each give up an electron to fill the sulfur atom's outer shell. Each of the sodium atoms are subsequently oxidized to form positive ions (Na^+), while the sulfur atom is reduced to a negative ion (S^{-2}). These oppositely charged ions combine to form sodium sulfide.

Acid-Base Reactions

An acid-base reaction occurs when a strong acid—a substance capable of donating hydrogen ions (H^+)—reacts with a strong base—a substance capable of accepting hydrogen ions. Acid-base reactions produce water and a salt. Salts are defined in chemistry as ionic compounds where the cation (positive ion) is not H^+ and the anion (negative ion) is not O_2^- or OH^-. For example, hydrochloric acid (HCl) reacts with sodium hydroxide (NaOH) to produce sodium chloride (NaCl), which is a salt compound, and water: HCl + NaOH $\rightarrow$ NaCl + H_2O.

Displacement Reactions

Displacement reactions cause elements to displace each other from a compound. For example, magnesium (Mg) displaces titanium (Ti) in the following reaction: 2Mg + $TiCl_4$ $\rightarrow$ Ti + $2MgCl_2$. Manufacturers use this particular displacement reaction to extract titanium, valued for its strength and light weight by the aerospace and other industries, from the compound titanium tetrachloride ($TiCl_4$).

Exchange Reactions

Exchange reactions are driven by compounds wanting to

exchange ions in order to form more stable products, namely acids or salts. For example, titanium tetrachloride ($TiCl_4$) reacts violently with water as $TiCl_4$ exchanges one titanium ion (Ti^{+4}) for every four hydrogen ions (H^+) in the water:

$$TiCl_{4(g)} + 2H_2O_{(g)} \rightarrow 4HCl_{(g)} + TiO_{2(s)}.$$

This reaction produces hydrochloric acid (HCl) and titanium dioxide (TiO_2). Titanium dioxide typically occurs in mineral deposits as an impure black substance. Chlorinating titanium oxides from mineral compounds produces $TiCl_4$. Because titanium dioxide is needed as the major pigment in white paint, manufacturers use the above exchange reaction to isolate titanium dioxide from $TiCl_4$.

Factors Influencing Reactions

Chemical reactions occur when certain physical and chemical factors make conditions energetically favourable for the reactants to combine into products. Some factors, such as the potential energy (stored energy) associated with the reactants, can trigger a spontaneous chemical reaction.

If the products have a higher level of entropy (disorder among the particles) than the reactants, this difference can also initiate a chemical reaction. External factors, such as heat or the presence of a catalyst (a substance that increases reaction rate without being chemically changed), can trigger or increase the rate of a reaction.

Exothermic and Endothermic Reactions

Chemical reactions can occur spontaneously if the reactants possess more potential energy (stored energy) than the products. This type of reaction occurs spontaneously because of the downhill energy path (from more potential energy to less).

These reactions are called exothermic (heat-producing) reactions, because potential energy is converted to heat as the reactions proceed. Conversely, endothermic (heat-absorbing) reactions do not occur spontaneously because of the uphill energy path that exists.

The products of endothermic reactions contain more

potential energy than the reactants. As a result, energy must be added to trigger an endothermic reaction.

Entropy

Entropy is the tendency for matter to become disordered. Nature requires the input of energy to maintain an ordered state—a bedroom will become messy if not periodically cleaned; a car will eventually fall into disrepair if not regularly serviced. Entropy is an important force in chemistry. If other factors influencing a reaction are held equal, a chemical reaction will proceed spontaneously if the products have higher entropy (are more disordered) than the reactants. This law explains why ozone (O_3) gas can spontaneously decompose into molecular oxygen (O_2): $2O_{3(g)} \rightarrow 3O_{2(g)}$. This reaction occurs because the molecular order is diminished, resulting in a higher level of entropy.

Heat

Applying heat to matter in the solid, liquid, or gas phase adds energy to the substance, causing the atoms to move faster and collide into each other with greater force. As a result, heat speeds up a chemical reaction by bringing atoms into contact with each other with greater force and frequency.

Catalysts

Catalysts are substances that trigger or speed up chemical reactions (without chemically altering the catalysts in the process). A catalyst combines with a reactant to form an intermediate compound that can more readily react with other reactants. An example of this is the formation of sulfur trioxide (SO_3), which is an important ingredient for producing sulfuric acid (H_2SO_4). Without a catalyst, sulfur trioxide is made by combining sulfur dioxide (SO_2) with molecular oxygen:

$$2SO_2 + O_2 \rightarrow 2SO_3.$$

Because this reaction proceeds very slowly, manufacturers use nitrogen dioxide (NO_2) as a catalyst to speed production of SO_3:

Step one: NO_2 (catalyst) + $SO_2 \rightarrow NO + SO_3$ (SO_3 is extracted and combined with steam to produce sulfuric acid)

Step two: NO (from Step One) + O_2 →NO_2 (catalyst that is reused in step one) In the above reactions, nitrogen dioxide (NO_2) acts as a catalyst by combining with sulfur dioxide (SO_2) to form both sulfur trioxide (SO_3) and nitrogen monoxide (NO). The sulfur trioxide is removed from the process (to be used in the production of sulfuric acid).

Nitrogen monoxide (NO) is subsequently combined with molecular oxygen (O_2) to produce the original catalyst, nitrogen dioxide (NO_2), which can be continually reused to catalyze sulfur trioxide (SO_3).

GENERAL TYPES

Combination

This is when two or more elements or compounds unite to form a single compound.

For example:

$2Mg + O_2 \rightarrow 2MgO$,

$KCN + S \rightarrow KSCN$,

$K_2O + SO_3 \rightarrow K_2SO_4$

Decomposition

This is when a compound breaks up into two or more elements or simpler compounds.

For example:

$2H_2O_2 \rightarrow 2H_2O + O_2$,

NH_4NO_3 – heat → $2H_2O + N_2O$,

$2FeSO_4$ – heat → $Fe_2O_3 + SO_3 + SO_2$

A reversible decomposition is called a "dissociation".

Examples:

$PCl_5 \rightleftharpoons PCl_3 + Cl_2$,

$2HI \rightleftharpoons H_2 + I_2$,

$NH_4Cl \rightleftharpoons NH_3 + HCl$

Displacement

This is when one element replaces another in a compound, or one compound replaces another in a larger compound.

For example:

$Fe_2O_3 + 2Al \rightarrow 2Fe + Al_2O_3,$

$Zn + CuSO_4 - aq \rightarrow Cu + ZnSO_4,$

$4HNO_3 + P_4O_{10} \rightarrow 4HPO_3 + 2N_2O_5,$

Double Decomposition

This is when two compounds interact by exchange of parts to produce two new compounds. The parts may be atoms or radicals.

Examples:

$AgNO_3 + NaCl - aq \rightarrow AgCl + NaNO_3,$

$PCl_3 + 3AgF \rightarrow PF_3 + 3AgCl,$

$SnCl_4 + 4PhMgCl \rightarrow SnPh_4 + 4MgCl_2,$

Addition

This is combination viewed from the point of view of one of the reactants, which has further atoms or groups of atoms added to it.

For example:

$C_2H_4 + Br_2 \rightarrow C_2H_4Br_2,$

$2PCl_3 + O_2 \rightarrow 2POCl_3,$

$CuSO_4 + 5H_2O \rightarrow CuSO_4 \cdot 5H_2O,$

In each example, the first reactant is said to have atoms or groups of atoms from the second added to it.

Substitution

This is displacement or double decomposition viewed from the point of view of one of the reactants, in which one atom or radical is replaced by another atom or radical.

For example:

$CH_4 + Cl_2 \rightarrow CH_3Cl + HCl,$

$C_6H_6 + HNO_3 \rightarrow C_6H_5NO_2 + H_2O$

These are double decompositions:

$CH_3\text{–}H + Cl\text{–}Cl \rightarrow CH_3\text{–}Cl + H\text{–}Cl,$

$C_6H_5\text{–}H + HO\text{–}NO_2 \rightarrow C_6H_5\text{–}NO_2 + H\text{–}OH,$

INSERTION

This is an addition reaction in which an atom or group

of atoms is inserted between two atoms initially bound together.

An example is the reaction,

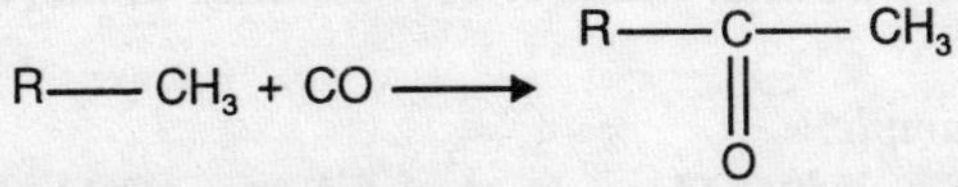

which takes place when R is one of a number of transition-metal radicals, e.g. $Mn(CO)_5$.

ISOMERIZATION

This is when a substance changes into another form with different properties but the same molecular formula.

For example:

CH_2=CH–CH_2–CH_3 – catalyst → CH_3–CH=CH–CH_3,

red HgI_2 – heat → yellow HgI_2,

The different forms are called "isomers" (Greek *isos*, equal), and the phenomenon of the existence of different forms "isomerism".

A rapidly reversible isomeric change is called a "tautomeric" change. The different forms are called "tautomers", and the phenomenon "tautomerism".

An isomeric change in the case of an element is called an "allotropic" change. The different forms are called "allotropes", and the phenomenon "allotropy".

Polymerization

This is when a substance changes into another substance with the same composition but a much higher molecular mass.

For example:

nC_2H_4 – heat, pressure, catalyst → $(–CH_2–CH_2–)_n$

The product of such a reaction is called a "polymer", and the starting material the corresponding "monomer" (Greek *polus*, much; *meros*, share; *monos*, alone).

The term polymerization is also used for processes in which a polymer is formed, not from the monomer, but from other reactants of low molecular mass. This usage is somewhat misleading, but is well established.

Oligomerization

This is similar to polymerization except that the product contains only a small number of monomer units (Greek *oligos*, small).

For example:

$$3C_2H_2 \text{ — heat, catalyst} \rightarrow C_6H_6$$

The degree of oligomerization is specified by the numerical prefixes di, tri, etc., as in dimer, trimerize, etc.

Some Special Types of Reaction

There are many special types of reaction. Among the more important are the following.

Precipitation

This is when an insoluble solid is formed in a reaction taking place in solution.

For example:

$NaCl + AgNO_3 \text{ — aq} \rightarrow AgCl\downarrow + NaNO_3,$

$K_2SO_4 + BaCl_2 \text{ — aq} \rightarrow BaSO_4\downarrow + 2KCl$

These are both double decompositions. Precipitation reactions are popular in analytical chemistry, providing both characteristic tests for different ions, and also, when the precipitate is sufficiently insoluble, a means of determining the amount ("gravimetric analysis").

Neutralization

This is a reaction between an acid and a base. It is discussed under "Acids, bases and salts".

Hydrolysis

This is a double decomposition involving water as one of the reactants. This splits into H and OH or 2H and O.

Examples:

$POCl_3 + 3H_2O \rightarrow 3HCl + PO(OH)_3,$

$CH_3CO_2CH_3 + H_2O \rightarrow CH_3CO_2H + CH_3OH,$

$SOCl_2 + H_2O \rightarrow 2HCl + SO_2,$

Redox Reactions

These are discussed separately under the heading "Oxidation and reduction".

Solvation

This is an addition reaction between a solute and a solvent.

For example:

$Na_2SO_4 + 10H_2O \rightarrow Na_2SO_4 \cdot 10H_2O$

$CuSO_4 + 5H_2O \rightarrow CuSO_4 \cdot 5H_2O$

$MgBr_2 + 2Et_2O \rightarrow MgBr_2 \cdot 2Et_2O$

The product of such a reaction is called a "solvate". If the solvent is water the adduct is called a "hydrate", and the process "hydration".

Complexation

This is an addition reaction that leads to the formation of a coordination entity.

CHEMICAL CHANGES

Chemical reactions, also called chemical changes, are not limited to happening in a chemistry lab. Here are some examples of chemical reactions with the corresponding chemical equations:

A silver spoon tarnishes: The silver reacts with sulfur in the air to make silver sulfide, the black material we call tarnish.

$$2\ Ag + S \rightarrow Ag2S$$

An iron bar rusts. The iron reacts with oxygen in the air to make rust.

$$4Fe + 3O2 \rightarrow 2Fe_2O_3$$

Methane burns. Methane combines with oxygen in the air to make carbon dioxide and water vapour.

$$CH_4 + 2O_2 \rightarrow CO_2 + 2H_2O$$

An antacid (calcium hydroxide) neutralizes stomach acid (hydrochloric acid).

$$Ca(OH)_2 + 2HCl \rightarrow CaCl_2 + 2H_2O$$

Glucose (simple sugar) ferments to ethyl alcohol and carbon dioxide. The sugar in grapes or from grain ferments with yeast to make the alcohol and carbon dioxide. The carbon dioxide is the gas that bubbles out of beer or champaign.

$$C_6H_{12}O_6 \text{ (glucose)} \rightarrow 2C3_2H_5OH \text{ (ethyl alcohol)} + 2CO_2$$

Alcohol plus oxygen becomes vinegar and a molecule of water. As in the fermentation of glucose, this is a more complex reaction than it appears here because it is a biochemical reaction.

$$C_2H_5OH + O_2 \rightarrow HC_2H_3O_2 + H_2O$$

As a general rule, biochemical happenings make poor examples of basic chemical reactions because the actual reaction is carried on within living things and under enzyme control.

PHYSICAL CHANGES

Here are some examples of changes that are NOT chemical reactions. In each case, the original material or materials may be reclaimed by physical processes.

- Water boils out of a kettle or condenses on a cold glass.
- An aluminum pot is put on a burner and gets hot.
- Dry ice goes from a solid to a gaseous form of carbon dioxide (*sublimation*).
- Gold melts or solidifies.
- Sand is mixed in with salt.
- A piece of chalk is ground to dust.
- Glass breaks.
- An iron rod gets magnetized.
- A lump of sugar dissolves in water.

CHEMICAL EQUATIONS AND REACTIONS

In order to write the chemical equations, you must first know the formulas for the materials involved. The formulas must be written on the proper side of the arrow - - reactants on the left and products on the right.

The order in which the reactants and products are written

does not matter, just as long as every material is on the proper side. Once the materials involved in the reaction are written correctly. If you need to draw a box around each participant in the reaction to keep your grubby paws off the materials, do it. Very often you will see the descriptions of the materials in the reaction in parentheses after the material. A gas is shown by (*g*). A solid material is shown by (*s*). A liquid is shown by (*l*). A material dissolved in water (an aqueous solution) is shown by (*aq*). An upwards pointing arrow (↑) indicates a gas being produced, and a downwards pointing arrow ,

$$[\downarrow]$$

indicates a solid precipitate being produced.

BALANCING EQUATIONS

Now comes the fun part, balancing the reaction. The Law of Conservation of Mass states that in a chemical reaction there is no loss of mass. Each type of element will have the same amount before the reaction and after the reaction, or as reactant and product.

But you cann't change the materials that participate in the reaction, so you must write an integer coefficient in front of (to the left of) each material in the reaction to make sure every type of atom has the same number on each side of the reaction. Lets start with the reaction of the Haber process:

Nitrogen gas plus hydrogen gas under pressure and at high temperature turn into ammonia. First write the materials correctly. Nitrogen and hydrogen are diatomic gases. Ammonia is a binary covalent memory item. The nitrogen and hydrogen are the reactants, and the ammonia is the product. Leave room for the coefficients in front of the materials.

$$N_2 + H_2 \rightarrow NH_3$$

You can begin with either the nitrogen or the hydrogen. There are two nitrogen atoms on the left and only one on the right. In order to balance the nitrogen atoms, place a '2' in front of the ammonia.

$$N_2 + H_2 \rightarrow 2NH_3$$

There are two hydrogens on the left and six on the right.

We balance the hydrogens by placing a '3' in front of the hydrogen gas.

$$N2 + 3H_2 \rightarrow 2NH_3$$

Now go back and check to make sure everything is balanced. There are two nitrogen and six hydrogens on both sides of the reaction. It is balanced. There is no coefficient shown in front of the nitrogen. There is no need to write ones as coefficients.

The reaction equation is:

$$N_2 + 3H_2 \rightarrow 2NH_3$$

BALANCING IONIC EQUATIONS

Silver nitrate and calcium chloride solutions combined produce a precipitate of silver chloride and leave a solution of calcium nitrate.

This time we have ionic compounds in the reaction. Until you are sure of the compounds, you might want to write the ionic materials as the ions, as demonstrated here.

$$Ag^+(NO_3)^- + Ca^{2+}Cl^-{}_2 \rightarrow Ag^+Cl^- + Ca^{2+}(NO_3)^-{}_2$$

Notice that from one side to the other there is no change in the nitrate ion. In this case you can count the nitrate ion as a whole rather than splitting it up into nitrogen and oxygen. Your thoughts might go this way: How many silvers on the right? One.

How many silvers on the left? One. They are the same. How many nitrates on the left? One. How many nitrates on the left? One. How many nitrates on the right? Two. We need to put a coefficient of two in front of the silver nitrate.

$$2AgNO_3 + CaCl_2 \rightarrow AgCl + Ca(NO_3)_2$$

This changes the balance of silvers, so we have to put a two in front of the silver chloride.

$$2AgNO_3 + CaCl_2 \rightarrow 2AgCl + Ca(NO_3)_2$$

Now let's check again. Two silvers on each side. Two nitrates on each side. One calcium on each side and two chlorides on both sides.

The balanced reaction is:

$$2AgNO_3 + CaCl_2 \rightarrow 2AgCl + Ca(NO_3)_2$$

BALANCING EQUATIONS WITH WATER

Sulfuric acid and potassium hydroxide neutralize each other to make water and potassium sulfate. Here is an acid-base neutralization. These make a salt and water.

$$H_2SO_4 + K^+(OH)^- \rightarrow K_{+2}(SO4)^{2+} + H_2O$$

The water is made from the hydrogen ion of the acid and the hydroxide ion of the base. Notice that it is a lot easier to understand how to balance the reaction if you write the water as if it were an ionic compound.

$$H_2(SO_4) + K^+(OH)^- \rightarrow K^+{}_2(SO_4)^{2+} + H^+(OH)^-$$

This is easier now because the hydrogen in the acid does not get confused with the hydrogen in the hydroxide of the base. Two hydrogens on each side. One sulfate on both sides. Two potassiums and two hydroxides on each side.

$$H_2(SO_4) + 2K(OH) \rightarrow K_2(SO_4) + 2H(OH)$$

The reaction is now balanced.

Next is an example of having to go around the equation again. Phosphoric acid and calcium hydroxide react to make water and calcium phosphate.

$$H_3PO_4 + Ca^{2+}(OH)^-{}_2 \rightarrow H^+(OH)^- + Ca^{2+}{}_3(PO4)^{3-}{}_2$$

First put a three on the water to balance the hydrogen in the phosphoric acid.

$$H_3PO_4 + Ca(OH)_2 \rightarrow 3H(OH) + Ca_3(PO4)_2$$

Now put a two on the phosphoric acid to balance the phosphate from the calcium phosphate.

$$2H_3PO_4 + Ca(OH)_2 \rightarrow 3H(OH) + Ca_3(PO4)_2$$

We have changed the amount of hydrogen ion, so we will have to change it on the right again.

$$2H_3PO_4 + Ca(OH)_2 \rightarrow 6H(OH) + Ca_3(PO4)_2$$

And change the coefficient in front of the Ca(OH)2 to match the calcium on the right side.

$$2H_3PO_4 + 3Ca(OH)_2 \rightarrow 6H(OH) + Ca_3(PO4)_2$$

Only now does the rest of the equation balance with six hydrogens, six hydroxides, two phosphates, and three calciums on each side.

BALANCING BURNING REACTIONS

Most burning reactions are the oxidation of a fuel material

with oxygen gas. Complete burning produces carbon dioxide from all the carbon in the fuel, water from the hydrogen in the fuel, and sulfur dioxide from any sulfur in the fuel. Methane burns in air to make carbon dioxide and water.

$$CH_4 + O_2 \rightarrow H_2O + CO_2$$

Easy. Put a two in front of the water to take care of all the hydrogens and a two in front of the oxygen. Anything you have to gather should be considered last.

$$CH_4 + O_2 \rightarrow 2H_2O + CO_2$$
$$CH_4 + 2O_2 \rightarrow 2H_2O + CO_2$$

What if the oxygen does not come out right? Lets consider the equation for the burning of butane, C_4H_{10}.

$$C_4H_{10} + O_2 \rightarrow CO_2 + H_2O$$

Insert the coefficients for carbon dioxide and water.

$$C_4H_{10} + O_2 \rightarrow 4CO_2 + 5H_2O$$

We now have two oxygens on the left and thirteen oxygens on the right. The real problem is that we must write the oxygen as a diatomic gas. The chemical equation is not any different from an algebraic equation in that you can multiply both sides by the same thing and not change the equation. Multiply both sides by two to get the following.

$$2C_4H_{10} + O_2 \rightarrow 8CO_2 + 10H_2O$$

Now the oxygens are easy to balance. There are twenty-six oxygens on the right, so the coefficient for the oxygen gas on the left must be thirteen.

$$2C_4H_{10} + 13\ O2 \rightarrow 8CO_2 + 10H_2O$$

Now it is correctly balanced. What if you finally balanced the same equation with:

$4C_4H_{10} + 26O_2 \rightarrow 16CO_2 + 20H_2O$

or

$6C_4H_{10} + 39O_2 \rightarrow 24CO_2 + 30H_2O$

Either equation is balanced, but not to the lowest integer. Algebraically you can divide these equations by two or three to get the lowest integer coefficients in front of all of the materials in the equation.

Now that we are complete pyromaniacs, let's try burning isopropyl alcohol, C_3H_7OH.

$$C_3H_7OH + O_2 \rightarrow CO_2 + H_2O$$

First take care of the carbon and hydrogen.

$$C_3H_7OH + O_2 \rightarrow 3CO_2 + 4H_2O$$

But again we come up with an oxygen problem. The same process works here. Multiply the whole equation (except oxygen) by two.

$$2C_3H_7OH + O_2 \rightarrow 6CO_2 + 8H_2O$$

Now the number nine fits in the oxygen coefficient. (Do you understand why?) The equation is balanced with six carbons, sixteen hydrogens, and twenty oxygens on each side.

$$2C_3H_7OH + 9O_2 \rightarrow 6CO_2 + 8H_2O$$

BALANCING BY OVERVIEW

Some equations are just mean, nasty, and rotten and defy your efforts to balance them. For some of these equations, a process I call overview is useful. Take as an example the smelting of magnetite, an iron ore.

$$Fe_3O_4 + CO \rightarrow CO_2 + Fe$$

Unless you just happen to hit it right, you are unlikely to balance this equation with the trial method.

The process overview shows that for each oxygen that the magnetite has, one carbon monoxide must turn to carbon dioxide. The carbon monoxide and carbon dioxide must have a coefficient that is four times the coefficient of the magnetite. Leave the magnetite coefficient and put a '4' in front of the carbon monoxide and carbon dioxide.

$$Fe_3O_4 + 4CO \rightarrow 4CO_2 + Fe$$

The carbon and oxygen is balanced, leaving only the iron to be balanced.

$$Fe_3O_4 + 4CO \rightarrow 4CO_2 + 3Fe$$

TYPES OF COMMON IONIC REACTION

Synthesis Reactions

- Also called Combination, Construction, or Composition Reactions

The title of this section contains four names for the same

type of reaction. Your text may use any of these. The first of the names and will use synthesis where your text may use one of the other words. The hallmark of a synthesis reaction is a single product. A synthesis reaction might be symbolized by:

$$A + B \rightarrow AB$$

Two materials, elements or compounds, come together to make a single product. Some examples of synthesis reactions are: Hydrogen gas and oxygen gas burn to produce water.

$$2H_2 + O_2 \rightarrow 2H_2O \text{ and}$$

sulfur trioxide reacts with water to make sulfuric acid.

$$H_2O + SO_3 \rightarrow H_2SO_4$$

What would you see in a test tube if you were witness to a synthesis reaction? You would see two different materials combine. A single new material appears.

DECOMPOSITION REACTIONS

- Also called Desynthesis, Decombination, or Deconstruction

Of the names for this type of reaction, The first. Mozart composed until age 35. After that, he decomposed. Yes, a decomposition is a coming apart. A single reactant comes apart into two or more products, symbolized by:

$$XZ \rightarrow X + Z$$

Some examples of decomposition reactions are: potassium chlorate when heated comes apart into oxygen gas and potassium chloride,

$$2KClO_3 \rightarrow 2KCl + 3O_2$$

and heating sodium bicarbonate releases water and carbon dioxide and sodium carbonate.

$$6NaHCO_3 \rightarrow 3Na_2CO_3 + 3H_2O + 3CO_2$$

In a test tube you would see a single material coming apart into more than one new material.

Single Replacement Reactions

- Also called single displacement, single substitution, or activity replacement

Here is an example of a single replacement reaction: silver nitrate solution has a piece of copper placed into it. The solution begins to turn blue and the copper seems to disappear. Instead, a silvery-white material appears.

$$2AgNO_3 + Cu \rightarrow Cu(NO3)2 + 2Ag$$

A solution of an ionic compound has available an element. The element replaces one of the ions in the solution and a new element appears from the ion in solution. This type of reaction is called a replacement because a free element replaces one of the ions in a compound.

There are two types of single replacement reactions, anionic and cationic. A cationic single replacement is what happened in the case of the silver being replaced by the copper in the above reaction because both the silver and the copper are only likely to make cations. An anionic single replacement is also possible. Into a potassium iodide solution chlorine gas is bubbled. The chlorine is used up and the solution turns purple-brown from the iodine. This is an example of an anionic single replacement reaction.

$$2KI + Cl_2 \rightarrow 2KCl + I_2$$

Could you start with copper II nitrate and silver metal and get silver nitrate and copper metal, or could you start with potassium chloride and iodine and get potassium iodide and chlorine? No. The reactions donot work that way. You can arrange cations or anions in a list of which ion will replace the next. This type of list is an activity series. The activity series of cation elements (metals) shows that gold is the least active metal. That should not be surprising, because gold does not tarnish. If we were to consider the Group 1 elements only on the activity list, lithium is the least active and francium is the most active, with each larger element being more active than the smaller one above it on the Periodic Chart.

On the other side of the chart we could consider an activity series for anions.

Taking just the halogens, the smallest halogen, fluorine is the most active. As the size of the halogen increases down the chart, the activity decreases. If an element is more active than the element of the same sign in an ionic solution, the more active element will replace it.

DOUBLE REPLACEMENT REACTIONS

- *Also called double displacement or metathesis*

Some texts refer to single and double replacement reactions as solution reactions or ion reactions. That is understandable, considering these are mostly done in solutions in which the major materials we would be considering are in ion form. There is some good reason to call double replacement reactions de-ionizing reactions because a pair of ions are taken from the solution in these reactions. Lets take an example.

$$AgNO_3 + KCl \rightarrow AgCl_{(s)} + KNO_3$$

The equation does not tell the whole story. Dissolved silver nitrate becomes a solution of silver ions and nitrate ions. Potassium chloride ionizes the same way. When the two solutions are added together, the silver ions and chloride ions find each other and become a solid precipitate. (They rain or drop out of the solution, this time as a solid.) Since silver chloride is insoluble in water, the ions take each other out of the solution.

$$Ag^+ + (NO3)^- + K^+ + Cl^- \rightarrow AgCl + K^+ + (NO_3)^-$$

Here is another way to take the ions out of solution. Hydrochloric acid and sodium hydroxide (acid and base) neutralize each other to make water and a salt. Again the solution of hydrochloric acid is a solution of hydrogen (hydronium ions in the acid and base section) and chloride ions. The other solution to add to it, sodium hydroxide, has sodium ions and hydroxide ions. The hydrogen and hydroxide ions take each other out of the solution by making a covalent compound (water).

$HCl + NaOH \rightarrow HOH + NaCl$ or

$H^+ + Cl^- + Na^+ + (OH)^- \rightarrow HOH + Na^+ + Cl^-$

One more way for the ions to be taken out of the water is for some of the ions to escape as a gas.

$CaCO_3 + 2HCl \rightarrow CaCl_2 + H_2O + CO_2$

$Ca^{2+} + (CO3)^{2-} + 2\,H^+ + 2Cl^- \rightarrow Ca^{2+} + 2Cl^- + H_2O + CO_2$

The carbonate and hydrogen ions became water and carbon dioxide. The carbon dioxide is lost as a gas to the ionic solution, so the equation can not go back. One way to consider double replacement reactions is as follows: Two solutions of ionic compounds are really just sets of dissolved ions, each solution with a positive and a negative ion material. The two are added together, forming a mixture of four ions.

If two of the ions can form:

- An insoluble material,
- A covalent material such as water, or a gas that can escape, it qualifies as a reaction.

Not all of the ions are really involved in the reaction. Those ions that remain in solution after the reaction has completed are called *spectator ions*, that is, they are not involved in the reaction.

To deny agriculture the use of these chemical tools would be to jeopardize our agricultural economy and an adequate, well-balanced food supply for the public. Uninhibited insects and plant diseases would largely eliminate the commercial production of such crops as apples, peaches, citrus fruits, tomatoes and potatoes, to mention only a few and would drastically curtail the production of many other major crops.

The claim is often made that without chemicals to protect our orchards and fields from pests, we could not possibly have a modern, productive system of agriculture. There is abundant evidence that our fruit, vegetables and many other staple crops cannot be produced economically, efficiently and in an adequate volume without chemical protection from insects, plant diseases, weeds and other pests, observes George C. Decker, a noted American entomologist.

Although no one would want to jeopardize our agricultural economy, still less deny the American people a well-balanced diet, a great deal of public and professional concern has been aroused by the presence of pesticidal residues in food. An insecticide should be "used as a stiletto rather than a scythe," A. W. A. Brown writes. "Ideally it should be employed to punch a particular hole in the food chain, which, if it cannot be filled in by [other species in the] neighboring elements of the biota, at least should be done at such a time of season that it does not seriously disturb the biota."

In practice, the trend has gone in the opposite direction. Insecticides are not only becoming more lethal; they are being used with less and less discrimination. Aside from the ecological implications of this trend, the growing toxicity and variety of insecticidal residues in food have become one of the major health problems of our time.

The large-scale application of insecticides to field crops did not get under way until the 1860's, when a large number of American farmers began to use paris green (copper acetoarsenite) against the Colorado potato beetle. The effectiveness of this insecticide encouraged chemists to develop a variety of other formulas. Most of the new insecticides were inorganic compounds—preparations of arsenic, mercury and sulfur—which normally remained on the crop after harvesting. The residues ranged in amount from mere traces to quantities large enough to be hazardous to the consumer.

Although the prewar era of inorganic insecticides has been described as pristine in comparison with the present era of organic preparations, the situation was far from ideal. In the 1930's one of the most widely used inorganic compounds, lead arsenate, combined a suspected carcinogen (arsenic) with a cumulative poison (lead). Calcium arsenate seriously damaged the soil, especially soils low in humus. Many areas of Washington and South Carolina were rendered unproductive as a result of being dusted with large amounts of arsenicals.

Harmful as they are, however, prior to 1945 inorganic insecticides were employed in relatively modest quantities and in a limited variety. The most lethal preparations were used primarily for agricultural purposes and the greater part of the residue could be removed from the surface of the crops by careful washing.

After World War II, the insecticide problem acquired entirely new dimensions. The total quantity of insecticides used prior to the war was but a fraction of the amount employed in the mid-1950's. Annual domestic sales of pesticides (primarily insecticides) increased more than sevenfold between 1940 and 1956. Inorganic insecticides, though still in use, declined markedly in importance; they were supplanted for the most part by a whole new group of organic preparations.

Of these the best known is DDT, a chlorinated hydrocarbon; it was synthesized as early as 1874, but it remained a -known chemical compound until the late 1930's, when the Swiss chemist Paul Müller discovered its insecticidal properties.

The research of Thomas and his colleagues led to the discovery of the insecticidal properties of lindane, a form of BHC. Significant work was also being done in Germany during this period. Shortly before the war, Gerhard Schrader, while engaged in research on poison gases, discovered a series of extremely lethal organic phosphates, from which parathion, malathion and other now familiar preparations were developed. The list continued to grow until an impressive number of chlorinated hydrocarbons and organic phosphates were available for use on farms and in gardens, warehouses, stores and homes.

In retrospect, the carelessness with which some of these insecticides were delivered to the public is incredible. The chemical behaviour of the chlorinated hydrocarbons inside the human body is still unknown, although it has been determined that both the chlorinated hydrocarbons and the organic phosphates are powerful nerve toxicants.

The organic phosphates are known to inactivate the enzyme cholinesterase, which participates in the chemical control of nerve impulses. Solutions of organic insecticides are absorbed through the skin as well as the lungs and digestive tract. The skin does not offer an effective barrier to the entry of these toxicants when they are used in liquid form.

The younger an organism, the more easily it is poisoned. Both groups of insecticides produce effects in warm-blooded animals very similar to those they produce in insects. Farm animals poisoned by DDT and other chlorinated hydrocarbons first become restless and excitable; the next stage is characterized by twitching, spasms and, finally, convulsions. Animals poisoned by the organic phosphates usually die of respiratory failure, but in these cases, too, convulsions have been known to occur.

Fortunately, DDT, the first of the new organic insecticides to be widely used, is one of the milder chlorinated hydrocarbons. The compound began to reach American farms in 1945.

As test samples of DDT had arrived in the United States only three years earlier, American research on the compound up to that time had been necessarily limited. Some of its more disquieting features were still under study when many Americans began using it with reckless abandon as a "magic bullet" against insect pests. At this stage, very little was known about the way the human body reacts to DDT, the possibility of chronic toxicity, the existence of the indirect routes by which DDT enters the human body and DDT's stability under varying agricultural conditions. By 1950 when a number of these questions had been answered, it had become evident that although the insecticide is certainly a "magic bullet," the target toward which it is traveling is very much in doubt.

It was found that DDT tends to concentrate heavily in fat. A relatively low dietary intake of the insecticide is magnified from six to twenty-eight times in the fat deposits of the body. According to A. J. Lehman, of the F.D.A., the continual ingestion of as little as 5 parts per million (ppm) of DDT in the diet produces "definite though minimal liver damage" in rats.

Such an intake is not likely to be uncommon in the United States. The F.D.A. has placed a residue tolerance of 7 parts per million of DDT on nearly every fruit, vegetable and portion of animal fat that reaches the American dinner table. But the agency does not know how many farmers are honoring these tolerances.

The reputation of DDT suffered its first blow only two years after the end of the war, As early as 1947, Italian entomologists noted that DDT-resistant flies were appearing in many parts of Italy. Similar reports began to come from California and other areas in the United States. By 1950, American surveys had shown that previously susceptible strains of flies "were becoming resistant to DDT over a period of 2 or 3 years."

Entomologists and public health workers were still hopeful that the malaria-carrying mosquito would continue to succumb to DDT, but this insect also developed DDT-resistant strains. As the number of resistant insects began to multiply, attention shifted to more lethal preparations, such as chlordane, dieldrin and the organic phosphates.

CHLORDANE AND DIELDRIN -ALDRIN GROUP

Chlordane and the dieldrin-aldrin group of preparations are highly toxic chlorinated hydrocarbons. A portion of the testimony and data that the Delaney Committee received on chlordane is summarized in its report as follows: "The Director of the Division of Pharmacology of the Food and Drug Administration testified that from a chronic viewpoint chlordane is four to five times more toxic than DDT and that he would hesitate to eat food that had any residue on it whatever.

Dieldrin and aldrin, according to Lehman, create a danger of chronic toxicity that "may be similar" to that posed by chlordane. These insecticides have the "rather rare property of being more toxic by skin absorption than by oral ingestion, the ratio being about 10:1. To state this another way, dieldrin and aldrin are approximately 10 times more poisonous when applied to the skin than when swallowed. Neither of these compounds is irritating to the skin in low concentrations so that the individual has no warning of skin contact."

The F.D.A. permits residues of chlordane, dieldrin and aldrin on nearly all the vegetables and fruits sold to American consumers. In view of current spraying practices, it is not unusual for foods to accumulate residues of several chlorinated hydrocarbons.

The fact that the various insecticides have different chemical properties is small reason for comfort. "Studies in our laboratories have shown that when low concentrations of several insecticides are fed to rats the effect is additive," Lehman writes.

"For example, when 2 ppm each of DDT, chlordane, lindane, toxaphene and methoxychlor, making a total of 10 ppm of chlorinated hydrocarbons, are added to the diet of rats, liver damage results. With the possible exception of chlordane, none of the insecticides listed, when fed individually at 2 ppm, would injure the liver, but in their combination their effect is additive." Where do we stand today in our chemical war against the insect world?

Although partial control has been achieved over some carriers of disease and other pests are no longer the nuisance they once were, the basic problems of insect infestation remain unsolved. Herbert F. Schoof observes that "man's insect enemies are adapting to successful coexistence with pesticides almost as rapidly as new technics and toxicants are produced." We are paying a heavy price for our questionable gains against agricultural pests. A larger quantity of insecticides is required almost every year.

The production of DDT in the United States nearly doubled between 1950, when the compound was no longer a novelty and 1958. Nearly 100 million pounds of aldrin, chlordane, dieldrin, endrin, heptachlor and toxaphene—chemicals virtually unknown to farmers in the 1940's—were produced in 1958. Approximately 3 billion pounds of pesticides, including herbicides, fungicides and fumigants, are produced annually in the United States. "The pesticide industry is growing by leaps and bounds and entomologists predict and chemical manufacturers hope for, a fourfold expansion in use of pesticides during the next ten to fifteen years," observes Clarence Cottam, of the Welder Wildlife Foundation.

"Today, well over 12,500 brandname formulations and more than 200 basic control compounds are on the market. Most of the currently used pesticides were unknown even ten years ago. Furthermore and contrary to the public interest, most new pesticides are decidedly more toxic, generally more stable and less specific in effect than those of but a few years back."

The postwar era also witnessed the introduction of highly potent chemical agents for increasing the weight of livestock. In 1947 the F.D.A. approved the use of stilbestrol pellets for "tenderizing" the meat of poultry. Stilbestrol, a synthetically prepared drug, causes bodily changes that are identical with those produced by certain female sex hormones (estrogens).

The drug was approved for use with the understanding that poultrymen would implant a single pellet in the upper region of the chicken's neck about a month or two before marketing.

Presumably this period was sufficient for the complete absorption of the chemical, so that no residues would be left in the flesh of the bird. Male birds treated with stilbestrol undergo radical physiological changes; their reproductive organs, combs and wattles shrivel and they lose all inclination to crow or fight. The drug tends to cause substantial increases in weight in both sexes, producing plump birds with skin that is noticeably smooth in texture.

Careful investigation has shown that stilbestrol increases the over-all fat content of chickens by 40 per cent. In the hearings before the Delaney Committee, Robert K. Enders, professor of zoology at Swarthmore College, expressed his agreement "with those endocrinologists who say that the use of the drug to fatten poultry is an economic fraud. Chicken feed is not saved; it is merely turned into fat instead of protein. Fat is abundant in the American diet so more is undesirable. Protein is what one wants from poultry.

By their own admission it is the improvement in appearance and increase in fat that makes it more profitable to the poultryman to use the drug. This fat is of very doubtful value and is in no way the dietary equal of the protein that the consumer thinks he is paying for." By the late 1950's approximately 100 million hormone-treated fowl were being marketed annually to American consumers. In a number of limited surveys, the F.D.A. found that stilbestrol pellets often remained in the necks of poultry until after the birds had reached wholesale outlets; the pellets were not completely absorbed.

Stilbestrol was detected in the edible tissues of hormone-treated fowl even after the heads and necks had been removed. Finally, in December 1959, the F.D.A. announced that, on its recommendation, the poultry industry had agreed to discontinue the use of the synthetic estrogen.

It is doubtful whether either the F.D.A. or the poultry industry had very much choice. Stilbestrol is suspected of being a cancer-promoting drug and recent federal legislation specifically prohibits the presence of carcinogenic additives in food. Pellets of stilbestrol are still implanted in the ears of cattle, however and the drug is widely used as a supplement to the feed of steers and sheep. The official opinion today is that no residues of the drug will remain in large animals if breeders obey F.D.A. regulations, but the evidence to support this opinion is open to question.

Index

G

H

I

L

M

N

O

P

Q

R

S

T

V